T0302236

Materials and Devices
for Thermal-to-Electric
Energy Conversion

MATERIALS RESEARCH SOCIETY
SYMPOSIUM PROCEEDINGS VOLUME 1166

Materials and Devices for Thermal-to-Electric Energy Conversion

Symposium held April 13–17, San Francisco, California, U.S.A.

EDITORS:

Jihui Yang
General Motors R&D Center
Warren, Michigan, U.S.A.

George S. Nolas
University of South Florida
Tampa, Florida, U.S.A.

Kunihito Koumoto
Nagoya University
Nagoya, Japan

Yuri Grin
Max-Planck-Institute for
Chemical Physics of Solids
Dresden, Germany

Materials Research Society
Warrendale, Pennsylvania

CAMBRIDGE
UNIVERSITY PRESS

University Printing House, Cambridge CB2 8BS, United Kingdom

One Liberty Plaza, 20th Floor, New York, NY 10006, USA

477 Williamstown Road, Port Melbourne, VIC 3207, Australia

314-321, 3rd Floor, Plot 3, Splendor Forum, Jasola District Centre, New Delhi - 110025, India

79 Anson Road, #06-04/06, Singapore 079906

Cambridge University Press is part of the University of Cambridge.

It furthers the University's mission by disseminating knowledge in the pursuit of education, learning and research at the highest international levels of excellence.

www.cambridge.org
Information on this title: www.cambridge.org/9781605111391

Materials Research Society
506 Keystone Drive, Warrendale, PA 15086
http://www.mrs.org

© Materials Research Society 2009

First published 2009
First paperback edition 2012

Single article reprints from this publication are available through University Microfilms Inc., 300 North Zeeb Road, Ann Arbor, MI 48106

CODEN: MRSPDH

A catalogue record for this publication is available from the British Library

ISBN 978-1-605-11139-1 Hardback
ISBN 978-1-107-40826-5 Paperback

CONTENTS

*Invited Paper

*Invited Paper

OXIDE MATERIALS

*Invited Paper

PREFACE

Symposium N, "Materials and Devices for Thermal-to-Electric Energy Conversion," held April 13–17 at the 2009 MRS Spring Meeting in San Francisco, California, was the ninth in a series of state-of-the-art materials and technologies for direct thermal-to-electric energy conversion symposia with primary focus on material and technological advances of thermoelectrics and thermionics [see MRS Symposium Proceedings Volumes 234, 478, 545, 626, 691, 793, 886 and 1044]. In this symposium there were 93 contributed presentations, including 17 invited talks and 28 poster presentations. These presentations showed the continuing technological development in thermal-to-electric energy conversion from research in academia, national laboratories, and industry in the United States, Asia and Europe. The symposium covered a broad range of topics in the areas of materials, devices, and applications. The tutorial session was led by Dr. Thierry Caillat of NASA Jet Propulsion Laboratory, Mr. Francis Stabler of Future Technologies, and Dr. Ryoji Funahashi of the National Institute of Advanced Industrial Science and Technology, Japan, and covered space, automotive, and industrial applications of thermoelectric technology. Harald Böttner (Fraunhofer Institute for Physical Measurement Techniques) began the symposium with an overview of the state-of-the-art in high-temperature thermoelectric materials. Dr. Böttner's talk covered materials cost, manufacturability, availability, etc., in addition to their thermoelectric properties, which provides important insights on material choices for technology development. The symposium session on nanocomposite materials included an overview given by Mildred Dresselhaus (MIT) on bulk nanostructured materials, one of the recent intensively investigated areas in the thermoelectric community. Peter Rogl (University of Vienna) gave an overview on the potential of inverse clathrates for thermoelectric applications. Continued interest in the area of thermoelectrics was evidenced by the excellent level of attendance throughout the symposium.

As with previous symposia in this series, there were a large number of graduate student presentations. This continues to be a focus of our symposium, emphasizing the strong interest from our future scientists in this field of materials research. With the generous support from our sponsors, the symposium organizers were able to give presentation awards to six students:

Oral Presentations

Reja Amatya (Massachusetts Institute of Technology) "Materials for Solar Thermoelectric Generators"

Sabah Bux (University of California, Los Angeles) "High Temperature Thermoelectric Properties of Nano-Bulk Silicon"

Steven N. Girard (Northwestern) "Investigation of Solid-State Immiscibility and Thermoelectric Properties of the System PbTe-SnTe-PbS"

Matthew Beekman (University of South Florida) "Preparation and Fundamental Properties of Clathrate-II Intermetallic Phases: Materials with Potential for Energy Conversion Applications"

Poster Presentations

Tomomi Okada (Tokyo University of Science) "Preparation of Delafossite $CuYO_2$ by Metal-citric Acid Complex Decomposition Method"

Takashi Nemoto (Tokyo University of Science) "Output Power Characteristics of Mg_2Si and the Fabrication of a Mg_2Si TE Module with a Unileg Structure"

The organizers are most grateful for the support of the U.S. Office of Naval Research and ULVAC Technologies Inc. This support funded the student awards and allowed for travel funds to help support our contributing presenters.

Jihui Yang
George S. Nolas
Kunihito Koumoto
Yuri Grin

June 2009

MATERIALS RESEARCH SOCIETY SYMPOSIUM PROCEEDINGS

MATERIALS RESEARCH SOCIETY SYMPOSIUM PROCEEDINGS

Prior Materials Research Society Symposium Proceedings available by contacting Materials Research Society

Applications and Devices

Mater. Res. Soc. Symp. Proc. Vol. 1166 © 2009 Materials Research Society   1166-N01-01

Thermoelectrics for High Temperatures - A Survey of State of the Art

Böttner H.

Fraunhofer-Institute Physical Measurement Techniques IPM, Department for Thermoelectric and
Integrated Sensor Systems, Heidenhofstrasse 8, 79110 Freiburg, Germany

ABSTRACT

A survey of state of the art of the development of high temperature materials is presented
and will be discussed in comparison to the situation in the 1990th. An attempt will be made to
assess the state of the art of the materials thermoelectric properties, their technical level, and
possible potential for standardized device technology. Also a first assessment based on current
commodity prices for some important thermoelectric compounds will be made.

As a roundup advantages and drawbacks for some classical and upcoming compounds
will be given. The main challenges, which will have to be overcome to finally enable
thermoelectric power generation as a recycling technology of "nomadic" energy, will be
summarized. As a result, thermoelectrics should play an important role in the field of green
energies.

INTRODUCTION

Energy is a scarce resource. Nevertheless, heat can be found escaping unused wherever
you look. Around 60 percent of all fossil primary energy is converted into unused waste heat.
Thermogenerators (TEGs) are known to be able to use those otherwise forever lost treasures of
our earth. This makes TEGs useful assistants in a process known as "energy harvesting". In
contrast to competitive heat converters like Stirling engines, thermoelectric generators function
without moving parts.

Converting car waste heat into electrical energy on a large scale is a realistic scenario and
was demonstrated by the preliminary system presented by e.g. BMW during summer 2008. Fuel
economy improvement of 5 - 8% for highway driving was claimed by BMW.

3

To enable this technology for exploiting waste heat and thus contribute to a more efficient utilization of natural resources, thermoelectric materials and standardized so called high temperature modules for temperature differences, 500°C or even more, are a prerequisite. They must be easily accessible like today's Bi_2Te_3-based standard modules. To achieve this goal much effort is under way worldwide. Only if highly efficient, cost-effective TEGs for high temperatures will be commonly available, waste heat in automobiles or in large-scale industrial plants, such as furnaces and refuse incinerators, can be economically converted into usable electrical energy.

A simple estimation highlights the high potential of TEGs: If 10% of the German car fleet, which comprises around 5 million cars, will be equipped with 1 KW generators, and assumed this generator will be active 200 hours per year, the energy recovered will be equal to about 1TWh. It should be mentioned that the US car fleet amounts to about 220 million. A typical nuclear plant like Philipsburg in Germany provides an output of ~6.6 TWh.

MATERIAL DEVELOPMENT FROM 1990 TILL NOWADAYS

As far as thermoelectric materials are concerned, up to about 1990 all applications were covered by three compound families: the V_2-VI_3 compounds, based mainly on Bi_2Te_3, the IV-VI-compounds based on PbTe and the IV-IV, the SiGe-alloys. Figure 1 reflects this situation in a ZT plot versus temperature [1]. The bars indicate the long lasting "thermoelectric limit" of ZT 1 and the cross over point of ZT versus T dependence of the V_2-VI_3 (Bi_2Te_3) and IV-VI (PbTe) compounds. This line divides, by the author's definition, the low temperature regime from the "higher" temperature regime just at 500 K, as this number is quite easy to memorize. The 500K border also represents approximately the maximum permanent "temperature of use" for commonly used thermoelectric devices based on V_2-VI_3 compounds.

Since 1990 material development focuses on two main approaches
⇒ better conversion efficiency caused by higher ZT-values and
⇒ materials usable for temperatures higher than typical as for V_2-VI_3 compounds

4

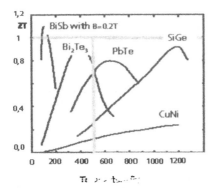

Figure 1. ZT versus temperature dependence for the main thermoelectric materials up to about 1990 [1], bars indicate the ZT=1 line and the border between low and high temperature material.

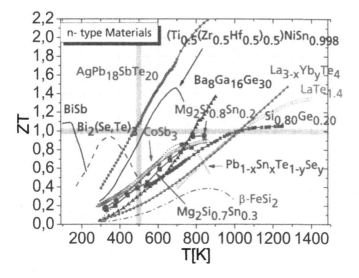

Figure 2. ZT versus temperature dependence for the main thermoelectric materials up to about July 2008, bars indicate the ZT=1 line and the border between low and high temperature material.

Figure 2 and figure 3 are representative for the state of the art for n- and p-type thermoelectric material, ~ July 2008.

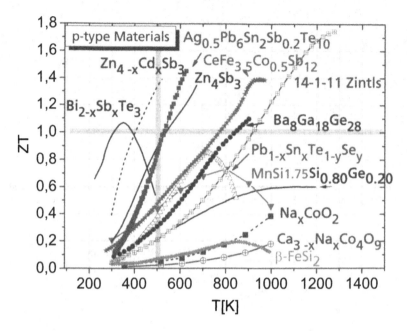

Figure 3. ZT versus temperature dependence for the main thermoelectric materials up to about July 2008, bars indicate the ZT=1 line and the border between low and high temperature material.

The progress is obvious. A huge number of new compound families have been investigated since and more or less all of them are still under development. It should be mentioned that since 1954 no new material was discovered in the low temperature range. For a better survey the compounds/compound families showing good ZT-values at temperatures ≥ 500 k are summarized in table 1. Effects on phonons reducing the thermal conductivity are in most cases responsible for the increase of the ZT-values. C. Godart [2] compiles typical reasons, the different effects on phonons y for nearly all mentioned "high temperature" materials, table 1.

Table I. Effects on phonons reducing the thermal conductivity according to C. Godart [2].

	Effects on phonons	Recent Materials
Complex structure	increase the number of optical phonon modes	Clathrate Chevrel Intermet. Yb$_{14}$MnSb$_{11}$ Skutterudite misfit oxides
Weakly bound atoms (or out of site positions) RGEC	increase disorder (rattling mode)	Skutterudite Clathrate Penta-telluride
Vacancies	increase disorder & mass fluctuations	Skutterudite half Heusler
Solid solutions	increase mass fluctuations	half Heusler-Mg$_2$(Sn,Si)
Impurities, inclusions	increase diffusion	new Bi$_2$Te$_3$TeCdSb PbTe -TAGS
Grain boundaries	reflect the mean free path of phonons	AgPb$_{18}$SbTe$_{20}$ nanocomposites

ECONOMIC ASPECTS OF "HIGH TEMPERATURE MATERIALS"

Cheap production of the thermoelectric materials in large (metric tons) quantities is a prerequisite for thermoelectric systems to enter mass markets. For a cheap production it is beneficial not to use rare and/or precious elements. Figure 4 shows how often these elements were used in the high temperature compounds families (indicated by black circles) and the relative abundance of chemical elements in the upper earth crust.

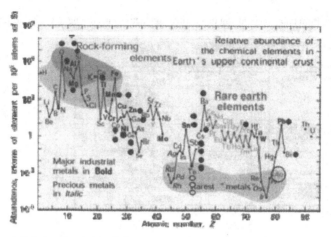

Figure 4. Relative abundance of elements in Earth's crust [3], "thermoelectric elements" are indicated by black circles.

Not all details are given but the two following information can be derived: from atomic number 8 (oxygen) till number 82 (bismuth) a lot of different elements are used for thermoelectric materials. Tellurium is approximately as rare as gold and therefore rather inappropriate for thermoelectric mass market applications. To get a better economic insight the price per kg thermoelectric material was calculated from 99.99% pure stock price elements (July 2008). In table 2 one may find the price in $/kg for the high temperature materials, compared to Bi_2Te_3, taking into account the element prices exclusively. The conclusion based on these economic estimations is obvious: for mass market high temperature materials the antimonide, silicides, scutterudites, Half Heusler and oxides seem to be well suited.

8

Table II. Price in $/kg for the high temperature materials, compared to Bi_2Te_3, taking into account the element prices exclusively.

Typ	Material	Price in $/kg (metals)
V-VI	Bi_2Te_3	140
IV-VI	$PbTe$	99
Zn_4Sb_3	Zn_4Sb_3	4
Silicides	$p\text{-}MnSi\,1.73$	24
	$n\text{-}Mg_2Si_{0.4}Sn_{0.6}$	18
	$Si_{0.8}Ge_{0.2}$	660
	$Si_{0.95}Ge_{0.05}$	270
Skutterutides	$CoSb_3$	11
Half-Heusler	$TiNiSn$	55
n/p-Clathrate	$Ba_8Ga_{16}Ge_{30}$	1000 without Ba
Oxides	$p\text{-}NaCo_2O_4$	17 without Na, O
Zintl Phasen	$p\text{-}Yb_{14}MnSb_{11}$	92
Th_3P_4	$La_{3x}Te_4$	160

TELLURIUM THERMOELECTRICS

Possible future effects on commodity prices are impressively illustrated by the "fever chart" of the tellurium price from April 2004 to April 2008, figure 5.

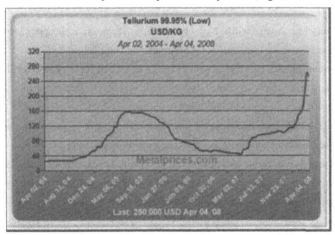

Figure 5. "Fever chart" of the tellurium price from April 2004 to April 2008 "

Taking into account the increasing market for CdTe-based solar cells, 1GW consume 100-200 metric tons Tellurium per year, and in addition the further main Te-consumption in the steel industry and tyre-production (Te is a vulcanizing agent), it can be presumed that the Te-price will be under speculation also in near term future. Furthermore the true tellurium consumption per year is unknown. Based on unconfirmed but plausible data around 20 million 4 x 4 cm^2 Bi_2Te_3-based modules were produced per year. Provided that each of these devices contains 10-20 gr. of Bi_2Te_3, ~100-200 metric tons tellurium will be consumed only for thermoelectric applications. Those data are not in line with the annual report of US geological commodity – Tellurium. For 2006, the US geological survey reported an overall refinery production of 128 metric tons.

A 4 x 4 cm^2 module may generate 10 Watts which equals ~0.5 W/cm^2. Thus 1,000 g converter material is necessary to generate 1 kW electrical energy. To equip 20 million cars with such a generator ~ 5.000 metric tons of tellurium would be needed. This exceeds the demand for the annual production of standard Bi_2Te_3 devices by 1.6 decades! For this reason a tellurium based thermoelectric mass market is inconceivable.

FAVORITE HIGH TEMPERATURE MATERIAL

The question which material will be best suited for high temperature application cannot be finally decided as of today. To demonstrate the opportunities of high temperature thermoelectric power generation on a limited basis PbTe will be the best choice. Just like PbTe (mainly due to economical reasons, see above) all other cheap materials have their specific disadvantages. It holds for any mentioned high temperature material, that no standardized commercially available modules exist. During the International Conference on Thermoelectrics 2004 [4], for instance, a silicide containing module was presented. However, up to now no reasonably priced product is available on the market. In the case of Half-Heusler alloys, the high ZT-values are waiting to be confirmed worldwide and the thermoelectric family is waiting for "engineering devices" to test modules containing Half-Heuslers. Oxides are very promising, taking into account the progress in ZT-values from 1997: ZT ~ 0.01 to ZT ≥ 0.3 – 0.4 nowadays.

For the oxides, however, the question of long-term stable high temperature Ohmic contacts arises.

SUMMARY

Under research are about a dozen so called high temperature material exhibiting ZT > 1. Enough ideas are underway for improvement in particular using nanoscale concepts combined with recently emerging densification technologies like Spark Plasma Sintering. Some preparation technologies are established and luckily the materials scientists are developing an increasing number of academic / experimental devices. The main challenges in the near and mid term future are: evaluation of air and chemical (ambient) stability, development of Ohmic contacts, large scale synthesis and first and foremost standardized commercial modules.

REFERENCES

1. C. Vining, European conference on thermoelectrics, Odessa, Ukraine, 10.-12. Sept. 2007, priv. communication
2. C. Godart APNFM, Dresden,Germany, 23-25.01.2008
3. http://en.wikipedia.org/wiki/Abundance_of_the_chemical_elements: relative abundance of elements in Earth's crust graphic from public domain source http://geopubs.wr.usgs.gov/factsheet/fs087-02/)
4. H. Kaibe, Proceedings 3rd International Conference on Thermoelectrics, Adelaide, Australia, 25-29 July, 2004, #36

Mater. Res. Soc. Symp. Proc. Vol. 1166 © 2009 Materials Research Society 1166-N01-06

Short Time Transient Behavior of SiGe-Based Microrefrigerators

Younes Ezzahri, James Christofferson, Kerry Maize and Ali Shakouri
Department of Electrical Engineering, University of California Santa Cruz
California, 95064-1077, USA.

ABSTRACT

We use a Thermoreflectance Thermal Imaging technique to study the transient cooling of SiGe-based microrefrigerators. Thermal imaging with submicron spatial resolution, 0.1C temperature resolution and 100 nanosecond temporal resolution is achieved. Transient temperature profiles of SiGe-based superlattice microrefrigerator devices of different sizes are obtained. The dynamic behavior of these microrefrigerators, show an interplay between Peltier and Joule effects. On the top surface of the device, Peltier cooling appears first with a time constant of about 10-30 microseconds, then Joule heating in the device starts taking over with a time constant of about 100-150 microseconds. The experimental results agree very well with the theoretical predictions based on Thermal Quadrupoles Method. The difference in the two time constants can be explained considering the thermal resistance and capacitance of the thin film. In addition this shows that the Joule heating at the top metal/semiconductor interface does not dominate the microrefrigerator performance or else we would have obtained the same time constants for the Peltier and Joule effects. Experimental results show that under high current values, pulse-operation the microrefrigerator device can provide cooling for about 30 microseconds, even though steady state measurements show heating. Temperature distribution on the metal leads connected to the microrefrigerator's cold junction show the interplay between Joule heating in the metal as well as heat conduction to the substrate. Modeling is used to study the effect of different physical and geometrical parameters of the device on its transient cooling. 3D geometry of heat and current flow in the device plays an important role. One of the goals is to maximize cooling over the shortest time scales.

INTRODUCTION

Much work has been conducted in the past to study the cooling performance of SiGe-based microrefrigerators both theoretically and experimentally in steady state or direct current (DC) and alternative current (AC) regimes [1-4]. Both superlattice based on Si/SiGe and also bulk SiGe thin film devices have been fabricated and characterized. Direct measurement of the cooling and cooling power density; along with material characterization have allowed extracting the key factors limiting the performance of these microrefrigerators [5]. SiGe-based microrefrigerators's cooling performance is based on the thermoelectric effect with thermionic enhancement. These devices can be monolithically integrated with Si microelectronics and optoelectronics in the IC industry. They can provide active cooling and offer an attractive, green and silent way to eliminate hot spots [5], thus the transient thermal behavior is important not only for improving device performance but to verify thermal models. It is well known that bulk material BiTe thermoelectric devices can have better cooling performance under pulsed mode operation [6], meaning that they can momentarily reach a colder temperature than when measured in the steady state. A similar behavior was predicted for the 3D SiGe-based thin films microrefrigerators devices. However limited metrology methods have prevented measurements.

EXPERIMENT

To investigate the transient thermal behavior of thin film superlattice microrefrigerators, we have used Thermoreflectance Thermal Imaging technique (TRI) [7]. TRI is one of the most sensitive thermal measurement techniques and offers much higher spatial, thermal and temporal resolution than infrared (IR) imaging. It is an optical non contact and non destructive tool for device thermal characterization that is based on the very small (0.01% per degree C) temperature dependence of the material reflection coefficient [7-9]. It is an active technique that uses light of the visible spectrum, usually a blue (~450nm) or green (~530nm) light emitting diode (LED), but in general can be optimized to the surface material of the device under test [7-9]. For thermal transient measurements, TRI is advantageous because it is an active technique. During the device's heating cycle, very bright, short light pulses are reflected off the device, and the duration of the LED light pulse is what determines the temporal image resolution.

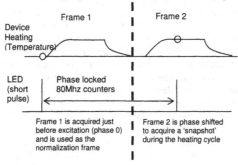

Figure 1: Transient thermal imaging scheme, illustrating timing of a short pulse LED for acquisition of a thermal *"snapshot"* of the device during the temperature cycle. By utilizing separate phase locked, programmable counters, different phases of the temperature cycle are obtained.

Using TRI, thermal transient information can be obtained using time domain or frequency domain techniques [8, 9]. The main challenge of such measurements is to resolve the small thermoreflectance signal among the various noise sources in the lab, and so extensive image averaging, and filtering is used. In this experiment, we have pursued a time domain, pulsed illumination technique, for several reasons, one primary reason being the simplicity in set-up, and lack of complex computer computation on the high-resolution CCD data, allowing for more averaging per unit time. Additionally, visualization of the time domain transient response is sometimes more intuitive, and results can be analyzed using Network Identification by Deconvolution (NID) method [10]. The latter is a powerful method to extract information about the various thermal resistances and capacitances of the device under study along the heat flux propagation path. Finally, by utilizing a high speed imaging laser diode that can be pulsed on the picosecond scale, the time domain transient TRI technique has the potential to provide ultra-short time scale (~ps) thermal images for characterization of material properties under excitation from a heating laser. The transient TRI experiment presented here utilizes a similar principle to the Pump-Probe Transient Thermoreflectance experiment performed with a femtosecond laser [11]. The primary difference being that in the case of transient TRI, the heating is created electrically rather than optically. In this experiment, the present limit for short time scale thermal imaging is 100ns, limited by the short pulse turn-on time of the LED used. At higher speeds, care must be taken to provide high-speed packaging for the LED, and additionally coaxial biasing probes must be used for device biasing.

Previous work on CCD based thermoreflectance transient measurements [12] used a very simple single pulse "*boxcar averaging technique*", which is sufficient to find transient information to the microsecond scale, however, as the time resolution increases (shorter light pulses) there is less light available on the CCD and the temperature sensitivity is decreased. The solution allowing for 100ns images, depicted in figure 1, is to provide many short light pulses for each CCD frame, and as long as we maintain precise timing (phase lock) between the device excitation pulse, and the LED illumination pulse, it becomes possible to obtain a single thermal "*snapshot*" of the device heating at a precise time in the heating cycle. The system was implemented using a combination of National Instruments programmable counter/time boards and TTL logic, necessary for the precise timing of signals. Because of the small thermoreflectance coefficient κ it becomes necessary to compare the value of the reflection coefficient at any time in the heating cycle to the value when the device is 'off' R_0, which removes the large DC term. At any time of the heating cycle, the CCD frames are averaged and differenced to obtain the change in the reflection coefficient ($\Delta R/R_0$) which is proportional to the change in the device surface temperature (ΔT) as function of time $\Delta R/R_0(t) = \kappa \times \Delta T(t)$. More technical details on the experiment can be found in the recent work of Christofferson et al [13].

THEORY

Modeling of the transient cooling performance of the 3D SiGe microrefrigerator is based on Thermal Quadrupoles Method (TQM) [14]. The TQM is a general analytical model that can be used to calculate electrical and thermal responses in a 3D geometry and in the AC regime, thus making it possible to distinguish, in some cases, the Peltier effect from the Joule effect. In the case of a pure sine wave electrical excitation, the Peltier effect appears at the same frequency as the operating current, whereas the Joule effect appears at the double frequency. The TQM is based on the solution of the one-dimensional Fourier's Heat Diffusion Equation in Laplace domain with a zero initial temperature. The TQM was used successfully to model the steady state and dynamic cooling performance of 3D SiGe based microrefrigerators [3, 4] and the results agree very well with thermocouple, and TRI measurements.

In the modeling of the transient cooling behavior, the microrefrigerator is assumed to be excited by a step electrical current and the top surface temperature variation as a function of time is calculated by taking into account all possible mechanisms of heat generation and conduction within the entire device. 3D heat and current spreading in the substrate is taken into account using analytic formulas. Besides, heat generation and conduction in the top metal lead of the device is also calculated using TQM. Because of the small temperature variations, all material properties are considered to be temperature independent [4]. In this paper, one will limit oneself

15

to only discuss the theoretical predictions in comparison with some experimental TRI results. The full detail of the application of TQM method to model the short time or transient cooling behavior of the 3D microrefrigerator will be presented in a future work.

DISCUSSION

Various samples ranging in size from 10x10μm² to 80x80 μm² were thermally imaged and compared to theoretical models. Figures 2(a-c) show the TRI results in comparison with the TQM simulation for two sizes 50x50μm² and 80x80μm².

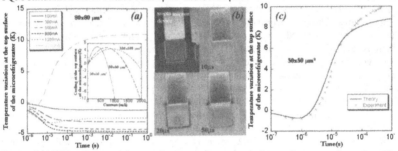

Figure 2: (a) Simulation of the transient temperature variation at the top surface of an 80x80μm² device for different electrical current amplitudes. The inset shows the steady state behavior for three different microrefrigerator sizes. (b) Image of the same device and transient thermal images as a response to a 1.7A, 166μs pulse. (c) Comparison between transient TRI result and TQM simulation for a 50x50μm² microrefrigerator obtained from 500ns to 500μs.

Figure 2(a) shows the simulation results using TQM, of the temperature variation at the top surface of an 80x80μm² device as a function of time, as a response to a step electrical current pulse of different amplitudes. The ohmic contact resistance at the top metal/semiconductor surface was taken to be $r_{oc}=10^{-6}\Omega.cm^2$ which is a realistic value for these devices [5]. For current amplitudes below the optimum value where the linear behavior of Peltier cooling dominates [inset of figure 2(a)], the temperature decreases as a function of time to stabilize in the cooling regime. On the other hand, when the current amplitude is higher than the optimum value, the temperature first decreases to reach a minimum and then it starts increasing. More interestingly is that the thin-film device can momentary cool below the ambient, even though the steady state shows heating [inset of figure 2(a)]. In figure 2(b), we can see that transient TRI results demonstrate clearly that we can capture the transient evolution of the microrefrigerator top surface performance as it goes from cooling to heating. The thermal images are taken on the top surface of an 80x80μm² SiGe based superlattice device and are acquired at 10μs, 20μs and 50μs. The frames are acquired every 10μs and the images are the response to a 1.7A, 166μs electrical excitation pulse [13]. These results confirm the simulation predictions in figure 2(a).

Figure 2(c) shows the results of transient TRI acquired from 500ns to 500μs on the top surface of a 50x50μm² SiGe based superlattice microrefrigerator and plotted in a logarithmic scale in comparison with TQM simulation. We have a satisfactory agreement between theory and experiment especially at short time scale. The dynamic behavior shows an interplay between Peltier and Joule effects. Peltier cooling is an interface effect located closer to the top surface of

16

the device. It appears first with a time constant of about 10-30μs. On the other hand Joule heating is a volume or bulk effect that takes certain time to reach the surface. Joule heating in the device starts taking over with a time constant of about 100-150μs. The difference in the two time constants can be explained considering the thermal resistance and capacitance of the thin film. In addition this shows that the Joule heating at the top metal/semiconductor interface does not dominate the microrefrigerator performance. If this was the case, we would have obtained the same time constants for Peltier and Joule effects.

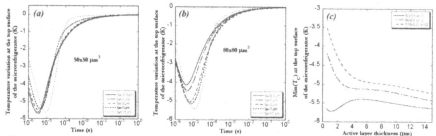

Figure 3: Variation of the transient temperature as a function of time at the top surface of a 50x50μm² (a) and an 80x80μm² (b) for different active layer thicknesses. (c) shows the variation of Min(T_C) as a function of the active layer thickness.

There are many material parameters that can be adjusted in the TQM simulations. Modeling is used to study the effect of different physical and geometrical parameters of the device on its transient cooling. Figures 3(a) and 3(b) show the transient cooling of a 50x50μm² and 80x80μm², respectively, from 1μs to 1s as a response to a step electrical current, calculated for different values of the thickness of the active SiGe layer t_{AL}. For this simulations, r_{oc} is taken to be $r_{oc}=0$. The amplitude of the electrical current is chosen to be the one for which the temperature at the top surface of the device vanishes at the steady state. We can see how changing t_{AL} affects the maximum cooling and the shape of the transient temperature over the time interval. This behavior is well captured in figure 3(c) that shows the variation of the minimum transient temperature or maximum transient cooling at the top surface of different device sizes over the same time interval, as a function of t_{AL}. Different behaviors occur depending on the microrefrigerator size. This effect is due to the 3D geometry of heat and current flow in the device that plays a very important role. The maximum cooling seems to manifest a slight minimum that is more pronounced for small sizes and disappears as the microrefrigerator size increases. Generally the maximum cooling increases by increasing t_{AL}. This behavior can be explained by considering the thermal capacitance of the active layer. The thermal capacitance increases by increasing t_{AL} and then it takes more time for Joule effect, which is a volume effect, to reach the top surface of the device where the temperature variation is considered. Since Peltier effect location does not change, this effect becomes more and more dominant as we increase t_{AL}.

For an 80x80μm²microrefrigerator size and fixed t_{AL}, we change respectively, the electrical conductivity σ_{AL}, the thermal conductivity β_{AL} and Seebeck coefficient S_{AL} of the active layer and we plot the maximum cooling and cooling power density over a time interval from 1μs to 1s. The curves are produced using the same conditions as in figure 3 above. As we can see in figures 4, both the maximum cooling and maximum cooling power density increase by increasing σ_{AL} and S_{AL}. On the other hand, the two of them decrease by increasing β_{AL}. As a mater of fact, increasing σ_{AL} and S_{AL} decreases Joule heating and increases Peltier cooling, respectively. On the other hand, increasing β_{AL} increases thermal conduction between the bottom and the top surface of the device where temperature variation and cooling power density are considered.

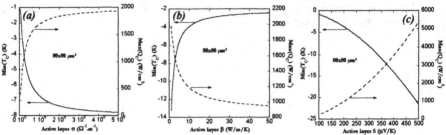

Figure 4: Variation of the transient maximum cooling and maximum cooling power density at the top surface of an 80x80μm² microrefrigerator over a time interval from 1μs to 1s as a function of the electrical conductivity σ_{AL} (a), thermal conductivity β_{AL} (b) and Seebeck coefficient S_{AL} (c) of the active SiGe layer.

Finally in figure 5, we show the behavior of the maximum cooling and maximum cooling power density at the top surface of the same 80x80μm² microrefrigerator device as a function of r_{oc}. By increasing r_{oc}, the Joule heating at the interface top metal/semiconductor, which is located on the same location as the Peltier cooling, increases. This effect decreases the cooling performance of the microrefrigerator and it appears to be one of the most key parameters. We have reached the same conclusion when dealing with the steady state behavior of microrefrigerators [4].

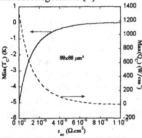

Figure 5: Variation of the transient maximum cooling and maximum cooling power density at the top surface of an 80x80μm² microrefrigerator over a time interval from 1μs to 1s as a function of the ohmic contact resistance at the top metal/semiconductor interface r_{oc}.

CONCLUSIONS

A CCD based transient Thermoreflectance Thermal Imaging system with 100ns temporal, 200nm spatial, 0.1C thermal resolution has been demonstrated. The thermal images can be acquired over 4 orders of magnitude, from 100 nanoseconds to 1 millisecond. The measured results on a SiGe based microrefrigerator device are in a satisfactory agreement with the TQM simulation based on a step function excitation type. Simulation and experiment show that the device can achieve a cooling below ambient temperatures under moderate currents for the first few tens of microseconds, even though steady state shows a net heating. This dynamic behavior shows an interplay between Peltier and Joule effects. Peltier cooling is an interface effect located closer to the top surface of the device. It appears first with a time constant of about 10-30μs. On the other hand Joule heating is a volume or bulk effect that takes certain time to reach the surface. Joule heating in the device starts taking over with a time constant of about 100-150μs. The difference in the two time constants can be explained considering the thermal resistance and capacitance of the thin film. We have considered only the variation of the active layer properties and the preliminary simulations results have shed some light on some of the key parameters. The top metal/semiconductor interface ohmic resistance seems to be one of the most dominant parameters. Further work planed is to consider the variation of the properties of the substrate and the top metal contact layer to investigate their effects on the total transient cooling performance of the microrefrigerator.

ACKNOWLEDGMENTS

The authors would like to knowledge the support of the Interconnect Focus Center, one of the five research centers funded under the Focus Center Research Program, a DARPA and Semiconductor Research Corporation Program.

REFERENCES

1. C. LaBounty, A. Shakouri and J. E. Bowers, J. Appl. Phys, **89**, (2001).
2. D. Vashaee, J. Christofferson, Y. Zhang, G. Zeng, C. LaBounty, X. Fun, J. Piprek, J. E. Bowers, E. Croke, and A. Shakouri, Microscale Thermophys. Eng, **9**, 99, (2005).
3. Y. Ezzahri, S. Dilhaire, L. D. Patiño-Lopez, S. Grauby, W. Claeys, Z. Bian, Y. Zhang and A. Shakouri, Superlattices and Microstructures, **41**, 7, (2007).
4. Y. Ezzahri, G. Zeng, K. Fukutani, Z. Bian and A. Shakouri, Microelectronics J, **39**, 981, (2008).
5. A. Shakouri, Proceeding of IEEE, **94**, 1613, (2006).
6. Q. Zhou, Z. Bian and A. Shakouri, J.Phys. D: Appl. Phys, **40**, 4376, (2007).
7. J. Christofferson and A. Shakouri, Rev. Sci. Instrum, **76**, 024903, (2005).
8. S. Grauby, S. Dilhaire, S. Jorez and W. Claeys, Rev. Sci. Instrum, **74**, 645, (2003).
9. P. L. Komarov, M. G. Burzo and P. E. Raad, Proceeding of THERMINIC 12, Nice, Côte d'Azur, France, September 27-29, (2006).
10. V. Székely and T. V. Bien, Solid-State Electronics, **31**, 1363, (1988).
11. Y. Ezzahri, S. Grauby, S. Dilhaire, J. M. Rampnoux and W. Claeys, J. Appl. Phys, **101**, 013705, (2007).
12. K. Maize, J. Christofferson and A. Shakouri, Proceedings of SEMITHERM 24, San Jose, California, USA, March 16-20, (2008).
13. J. Christofferson, Y. Ezzahri, K. Maize and A. Shakouri, Proceeding of SEMITHERM 25, San Jose, California, USA, March 15-19, (2009).
14. D. Maillet, S. André, J. C. Batsale, A. Degiovanni, and C. Moyne,, *"THERMAL QUADRUPOLES: Solving the Heat Equation through Integral Transforms"*, John Wiley & Sons, (2000).

Mater. Res. Soc. Symp. Proc. Vol. 1166 © 2009 Materials Research Society   1166-N03-14

Long-Term Performance of a Commercial Thermoelectric Power Generator

Euripides Hatzikraniotis[1], Konstantinos Zorbas[1], Theodora Kyratsi[2]
and Konstantinos M Paraskevopoulos[1]

[1] Department of Physics, Aristotle University of Thessaloniki, 54124 Thessaloniki, Greece
[2] Department of Mechanical and Manufacturing Engineering, University of Cyprus,
Nicosia, Cyprus

ABSTRACT

In this work, thermoelectric device was made, using a commercially available ThermoElectric Generator (TEG), in order to measure the gained power and efficiency for long-term performance. The module was subjected to sequential hot side heating at 200°C (392 °F), and cooling for 6000 cycles, in order to measure the TEG's power and EMF change. A 14% increase in the TEG's material resistance was found, as well as a 5% reduction in the Seebeck coefficient. After the experiment, the module was disassembled and thermoelectric p- and n-legs were examined using IR spectroscopy.

INTRODUCTION

Thermoelectric generators (TEG) make use of the Seebeck effect in semiconductors for the direct conversion of heat into electrical energy, which is of particular interest for systems of highest reliability or for waste heat recovery [1]. A generator usually consists of several pairs of alternating p- and n-type semiconductor blocks (generator legs), which are arranged thermally in parallel and connected electrically in a series circuit. Heating one side of the arrangement, the opposite side being cooled, induces the heat flow, which is partly converted into electrical power.

The possible use of a device consisting of numerous TEG modules in the wasted heat recovery of an internal combustion (IC) engine can considerably help the world effort for energy savings. Generally, the wasted heat from IC engines is a great percentage of the fuel's energy. In gasoline fuelled IC engines, about 75% of the total energy of the fuel is rejected in the environment [2]. The recovery of a 6% of the exhaust's energy could lead to 10% saving of fuel. Furthermore, the temperatures developed vary from high (about 900°C/1650°F at exhaust manifold) to medium (about 100°C/210°F in the engine coolant fluid) and thus the efficiency of the thermoelectric elements could be remarkable. The use of a thermoelectric generator device will offload the alternator and thus will reduce its size. However, temperatures much higher than the desired operating range could cause structural failure of the thermoelectric elements.

The requirements in the design of a system to be installed in a typical modern car are very high. The system should be able to operate reliably, with no or minimal maintenance requirements for at least 10 years (or 300,000 Km /186,411 miles) at external environment temperatures that could range from -40 to 50 °C (-40 to 122 °F) and occasionally in high humidity conditions. It should also be resistant to vibrations and strong mechanical stress and be as small as possible in weight and volume. In particular, a thermoelectric device to be placed on the exhaust, should be resistant to thermal shocks during the engine starting as the exhaust temperature rises from ambient temperature to 400 - 600 °C (750 to 1110 °F) at a time less than one minute. Furthermore, it should be resistant to very high temperatures (up to 900 ^{0}C / 1650 ^{0}F, depending on the type of engine and on the location of the device [2]), which could be developed for short or extended periods of time at different possible operating situations. Moreover, the performance must remain stable, despite the constantly changing conditions. Concurrently, according to the foregoing requirements in today's highly competitive automotive industry, the design should satisfy the low cost factor.

The exposure of thermoelectric materials to such environments causes, generally, an increase in electrical resistivity, as well as a decrease in material "figure of merit" Z. Furthermore, the continuous thermal cycling charge, could lead to reduced efficiency and lifetime of the TEG.

In this work, a thermoelectric device using a commercially available TEG was made, in order to measure the gained power and efficiency. As the TEG's reliability is a very important factor in waste heat applications, the module was subjected to sequential hot side heating and cooling for 6000 cycles, in order to measure the TEG's power and emf change. After the experiment, the module distortions were checked over and the changes in thermoelectric material were examined with the use of a Fourier transform infrared spectroscopy (FTIR).

EXPERIMENTAL

Commercial 2.5x2.5 cm Bi_2Te_3 modules with N=31 thermocouples (Melcor HT9-3-25) were used. The heater was made of copper and was attached directly to the top of the TEG module. In order to keep a constant cooling temperature, a liquid heat exchanger was used as cooler. All pieces were bonded together with two bolts at a pressure on TEG's surfaces of 4 MPa. In order to reduce the thermal contact resistance, all surfaces were lapped at a maximum roughness of about 25 μm and a thin layer of graphite thermal grease (Melcor GRF-159) was used.

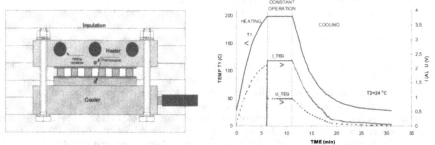

Figure 1: (a) Experimental setup and (b) Typical cooling-heating cycle

For temperature monitoring two K-type thermocouples were mounted, one in a hole of 1mm diameter near the bottom surface of the heater and the other in a thin copper plate on the bottom of the module, as shown in Figure 1a. An Eliwell EWTR 910 temperature controller controlled the temperature of the heater. For the electrical measurements, an Agilent 34401A and a Metrahit multimeters were used as voltmeter and ammeter respectively.

In order to perform the reliability test, a Siemens Logo RC230 PLC was used for controlling the repeated heating-cooling typical cycle. Heating was done at hot side temperature of 200°C and cooling to RT. A typical heating-cooling cycle is shown in Figure 1b. An ohmic resistance of 0.45 Ohm (as near as the value of the internal TEG's resistance) was used as load, in order to achieve the maximum power.

Furthermore, in order to have an independent measure of material degradation, a "Z-meter" device, type DX4065 of RMT Ltd [3], was used. With this device we could measure the material "Figure of merit" Z, the TEG's electrical resistance R_{TEG} and the time constant τ (which defines the time period necessary for a module to reach the steady state in response to the switching of the current) at an average temperature of 20°C.

RESULTS and DISCUSSION

The TEG module was subjected to sequential 6000 heating-cooling cycles, each one lasting about 30minutes. During the heating phase the external ohmic load was disconnected for quick reaching the desired temperature, while the load remained connected during the cooling phase.

A model for the efficiency test

Four basic physical phenomena are associated with the operation of thermoelectric generators (TEG), namely, the Seebeck effect, the Peltier effect, the Thomson effect and the Joule effect. Under steady state conditions, and neglecting the contribution of Thomson effect as small, the equation that governs the heat flow:

at the hot side: $\qquad Q_H = K_{TEG} \cdot (T_H - T_C) + S_{TEG} \cdot T_H \cdot I - \frac{1}{2} I^2 R_{TEG}$ (1)

at the cold side: $\qquad Q_C = K_{TEG} \cdot (T_H - T_C) + S_{TEG} \cdot T_C \cdot I + \frac{1}{2} I^2 R_{TEG}$ (2)

net power produced $\qquad P_{TEG} = Q_H - Q_C = S_{TEG} \cdot (T_H - T_C) \cdot I - I^2 R_{TEG}$ (3)

voltage produced $\qquad V_{TEG} = S_{TEG}(T_H - T_C) - IR_{TEG}$ (4)

where K_{TEG} is the total thermal conductance of the N couples, S_{TEG} is the total Seebeck coefficient , R_{TEG} is the total resistance. Therefore, the P_{TEG}, Q_H and Q_C can easily be calculated, if the material properties are known. In practice, it is difficult to measure the temperature of both the hot and the cold junction (T_H and T_C), as the p- and n-legs are interconnected by metal (typically Cu) and are thermally in parallel between two ceramic plates. However, it is feasible to measure the temperatures T_1 (hot side area) and T_2 (cold side area) at some distance of the ceramic plates, and taking into account the thermal resistances of the copper, the ceramic plate layers, and the thermal contact resistance between the heater and the ceramic plate, the hot and the cold junction temperatures (T_H and T_C) can be evaluated [4]. The calculation of T_H and T_C assumes that the average values of the material's Seebeck coefficient α, resistivity ρ and thermal conductivity κ, for the p- or the n- leg, are known, or can be measured independently [5].

Reliability test

In Fig.2a it can be seen that after the first 50 cycles the module reveals a power drop of about 12.5% and an EMF drop of about 15%. This abrupt change can be attributed to the deterioration of the thermal grease, caused by the elevated temperature. From our theoretical model, we can estimate an increase in the hot side thermal resistance from 0.1486 to 0.3 °K/W. This means that the temperature difference between the TEG's hot and cold side T_H-T_C, decreases from 164 to 158K and the average temperature from 385 to 381K.

In Fig. 2b are shown the evaluated change in average (p- and n-) material resistivity (ρ) and the Seebeck coefficient (α) during the reliability test. TEG's resistance R_{TEG} (and

23

consequently material resistivity ρ) was evaluated from the slope of the linear trend ($V_{TEG}=U_0-I*R_{TEG}$), where U_0 is the open circuit voltage of the TEG when maximum temperature is reached, V_{TEG} is the voltage after the load is connected, and I is the current that passes. As U_0 is proportional to Seebeck coefficient α, the change in material Seebeck coefficient can be also evaluated [6].

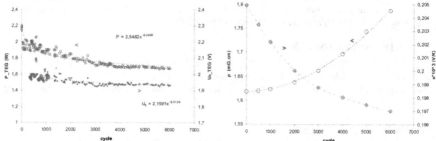

Figure 2: (a) TEG's maximum gained power and EMF during reliability test
(b) Change in material resistivity and Seebeck coefficient during the reliability test.

The measurements with the Z-meter show an initial TEG resistance of 0.31 Ohm (for an average temperature 20°C), since the resistance after the 6000 cycles is 0.36 Ohm, or an increase of 13.9%. This value is in good agreement with our measurements in TEG resistance (for average temperature 108°C) of 10.4% increase. The material figure of merit Z (for an average temperature 20°C) initially was $2.474 \cdot 10^{-3}$ (K^{-1}) while after the 6000 cycles was $2.112 \cdot 10^{-3}$ (K^{-1}), which indicates a slight change in the thermal conductivity (<2%). This is verified by the the variation in the time constant of the TEG, which is conversely proportional to the material thermal conductance κ [7], had no significant change during the test.

The reduction in the Seebeck coefficient and the increase in resistivity, are indicative of material deterioration caused by the repeated thermal shock. The reduction in the Seebeck coefficient could be possibly attributed to inter-diffusion of Cu from the connecting metal plates to the p- or n-types of blocks. Consequently, the increase in resistivity could be attributed to microcracks formation, caused by the elevated temperature.

IR reflectivity
Infrared spectra were carried out on disassembled p- and n-type legs using a 113v Bruker spectrometer. The reflection coefficient was determined by typical sample-in, sample-out method with a gold mirror as the reference. Infrared measurements were taken at near normal incidence, at the frequency range from of 70cm^{-1} to 2000cm^{-1} with resolution of 2 cm^{-1}. As optimization for p- and n- type material is usually made by solid solutions of Bi_2Te_3 with Bi_2Se_3 (for the n-type) and with Sb_2Te_3 (for the p-type), IR spectroscopy results for the p- and n- type legs are compared to a freshly cleaved surface of crystalline Bi_2Te_3 and Bi_2Se_3 samples prepared from melt. Spectra were analyzed by Kramers-Kronig method; Kramers-Kronig algorithm was optimized for increased accuracy in the high- and low-end spectral regions.

The reference Bi_2Te_3 and Bi_2Se_3 samples show a high reflectivity at low wavenumbers and a well-defined minimum, at 245cm^{-1} and 756cm^{-1}, respectively, which is typical for Drude-type reflection. Some small structure is observed at the wavenumbers ~100 cm^{-1}. The n- and p-type legs do not show a typical Drude-type reflectivity, however, as both n- and p-type legs are pressed polycrystalline samples, a significant surface roughness is expected, and this is manifested by a lowering trend in the reflectivity, at higher frequencies.

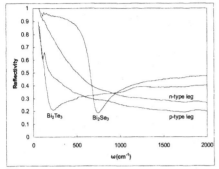 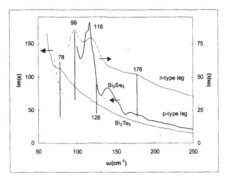

Figure 3: (a) IR reflectivity for n-type and p-type legs along with a reference Bi_2Se_3 and Bi_2Te_3 (b) Imaginary part of the dielectric function for n-type and p-type legs and a reference samples.

The IR reflectivity (R) is given by expression:

$$R(\omega) = \left| \frac{\sqrt{\varepsilon(\omega)} - 1}{\sqrt{\varepsilon(\omega)} + 1} \right|^2 \qquad (5)$$

where, $\varepsilon(\omega)$ is the complex dielectric function ($\varepsilon = \varepsilon_1 + i\cdot\varepsilon_2$), which, in the classical dispersion theory is expressed in terms of the IR-active modes and in addition a Drude contribution, to encounter for the free carrier plasma contribution to reflectivity. Thus, the dielectric function is related to the transverse (TO) optical mode frequency (ω_T) and the plasma frequency (ω_P) as:

$$\varepsilon(\omega) = \varepsilon_\infty + \sum \frac{\Delta\varepsilon_j \omega_{T,j}^2}{\omega_{T,j}^2 - \omega^2 + i\cdot\gamma_{T,j}\omega} - \frac{\varepsilon_\infty \omega_P^2}{\omega^2 + i\cdot\gamma_P\omega} \qquad (6)$$

where $\Delta\varepsilon_j$ and $\gamma_{T,j}$ is the oscillator strength and the damping constant for the j-th oscillator, γ_P is the damping constants for the plasmon, and ε_∞ is the high frequency value for the dielectric function. In Figure 3b is presented the imaginary part of the dielectric function, as derived by Kramers-Kronig calculation. As can be seen, in all cases, peaks are developed on top of an almost exponential decaying curve; free carrier contribution becomes evident from the rising trend in the $Im(\varepsilon)$ graph at low frequencies. Despite the fact that in the reference samples peaks are higher in value, which is indicative of a much lower damping factor, peak position is almost the same in both n-type and p-type samples as well as in reference ones. Analysis results are summarized in Table I. In addition to a Drude type reflectivity, two oscillators at 78cm^{-1} and 97cm^{-1} characterize the Bi_2Te_3 reference sample. Similarly, two oscillators at 117cm^{-1} and 178cm^{-1} (along with a Drude component) manifest the IR spectra for Bi_2Se_3 reference sample. In both the p- and n- type legs, the Drude component is present. In n-type leg, we find the oscillators of Bi_2Te_3 and, in addition the ones for Bi_2Se_3. In the p-legs the modes at 78cm^{-1} and

25

$99 cm^{-1}$ are attributed to Bi_2Te_3, while the additional mode at $128 cm^{-1}$, should be presumably attributed to Sb_2Te_3.

Table I: Spectral position of mail extrema in the experimental dependences of Im(ε) for the reference sample, the n-type and the p-type legs

sample	j=1 (cm^{-1})	j=2 (cm^{-1})	j=3 (cm^{-1})	j=4 (cm^{-1})	j=5 (cm^{-1})
Bi_2Te_3	78	97	-	-	-
Bi_2Se_3	-	-	117	-	178
n-type leg	?	99	118	-	176
p-type leg	79	99	-	128	-

CONCLUSIONS

In this work it is investigated the long-term performance and stability of a commercially available TEG under temperature and power cycling. With the development of a measuring device and a theoretical model, which takes into account the contact thermal resistances, the changes in resistivity and Seebeck coefficient of the thermoelectric material are evaluated. It is observed a 14% increase in the TEG resistance and a 5% reduction in the Seebeck coefficient. From the IR reflectivity measurements, it was verified that p- and n-legs are characterized by the modes of Bi_2Te_3, and Bi_2Se_3 or Sb_2Te_3.

ACKNOWLEDGMENTS

It is acknowledged the financial support of the project entitled "Application of Advanced Materials Thermoelectric Technology in the Recovery of Wasted Heat from automobile exhaust systems" by the Greek Secretariat of Research and Development under the bilateral framework with Non-European countries (Greece-USA).

REFERENCES

1. Rowe D. M. *CRC Handbook of Thermoelectrics*. London: CRC Press; (1994)
2. F. Stabler. *Automotive applications for high efficiency thermoelectrics*. DARPA/ONR Program Review and DOE High Efficiency Thermoelectric Workshop, San Diego, CA, March 24 – 27 (2002); F. Stabler, *Commercialization of Thermoelectric Technology*, Mater. Res. Soc. Symp. Proc., vol. **886**, p. 0886-F01-04.1 (2006)
3. G. Gromov, D. Kondratiev, A. Rogov, L. Yershova. *Z-meter: Easy-to-use Application and Theory*. RMT Ltd, Moscow.
4. K. Zorbas, E. Hatzikraniotis, K.M. Paraskevopoulos, *Power and Efficiency calculation and evaluation of material properties in Thermoelectric Power Generators*, Mater. Res. Soc. Proc. vol. **1044**, p.1044-U09-15 (2008)
5. E. Hatzikraniotis, K.T. Zorbas, K.M. Paraskevopoulos, *Material degradation of Thermoelectric Power Generators (TEG) for use in automobile*, presented in EMRS 2008 Spring Meeting, (symposium M: Unconventional thermoelectrics) Strasbourg (France), May 26-30, 2008 (Appl. Energy, submitted 2009)
6. K. Zorbas, E. Hatzikraniotis and K.M. Paraskevopoulos in *Study of Power efficiency in Thermoelectric Power Generators*, Proc. XXII Greek Conference of Solid State Physics and Science of Materials, Patras, Greece, September 24-27 (2006)
7. L.B. Yershova, G.G. Gromov I.A.Drabkin. *Complex Express TEC Testing*. RMT Ltd, Moscow

Nanocomposite and
Nanostructured Materials

Mater. Res. Soc. Symp. Proc. Vol. 1166 © 2009 Materials Research Society   1166-N02-01

The Promise of Nanocomposite Thermoelectric Materials

M.S. Dresselhaus[1], G. Chen[1], Z.F. Ren[2], K. McEnaney[1], G. Dresselhaus[1], and J.-P. Fleurial[3]
[1]Massachusetts Institute of Technology, 77 Massachusetts Ave., Cambridge, MA 02139 U.S.A.
[2]Boston College, 140 Commonwealth Ave., Chestnut Hill, MA 02467 USA
[3]Jet Propulsion Laboratory, 4800 Oak Grove Dr., Pasadena, CA 91109 USA

ABSTRACT

The concept of using nanocomposite thermoelectric materials in bulk form for practical applications is presented. Laboratory studies have shown the possibilities of nanostructures to yield large reductions in the thermal conductivity while at the same time increasing the power factor. Theoretical studies have suggested that structural ordering in nano-systems is not necessary for the enhancement of ZT, leading to the idea of using nanocomposites as a practical scale-up technology for making bulk thermoelectric materials with enhanced ZT values. Specific examples are presented of nanocomposite thermoelectric materials developed by our group based on the familiar silicon germanium system, showing enhanced thermoelectric performance through nano-structuring.

INTRODUCTION: THE ENERGY CHALLENGE

Thermo-electricity in the larger energy picture

During the 21st century inhabitants on our planet expect to be making a transition from today's fossil fuel based economy to an energy economy that will be sustainable over the long term. The drivers for this transition to a sustainable energy scenario include the increase in global population, going from today's 6.5 billion people to 8.9 billion (see figure 1(a)) by 2050. The last 30 years have not only witnessed an increasing energy demand per person, but people worldwide have aspired to a higher standard of living, resulting in a super-linear increase in energy demand over this time period (see figure 1(b)). Despite the large efforts in increasing energy efficiency and to some degree energy conservation, the increasing energy demands, especially from the developing world (see figure 1(b)), have in large part contributed to the super-linear aspect of the increasing energy demand worldwide. Extrapolating into the future, the energy demands are expected to double from today's 15±2 TW level to about the 30 TW level by mid-century. To address this huge increase in energy demand over a short time period of 40 years is a daunting challenge to our global population, not only from an energy standpoint, but also from an environmental standpoint as over the last thousand years there has been a notable increase in global both mean temperature (by ~1°C) and CO_2 levels (to about 380 PPM).

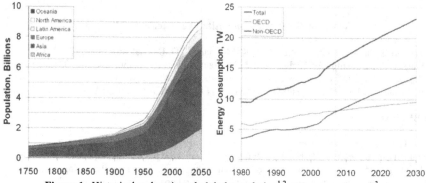

Figure 1. Historical and projected global population[1,2] and energy demand[3].

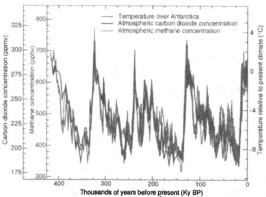

Figure 2. Ice core data[5] for temperature and atmospheric CO_2 and CH_4.

When looking at data from ice core records that go back in history, we see that natural events on our planet (occurring over a period of ~400,000 years) result in fluctuations in the temperature of about 10°C over 100,000 year periods, so that the man-made effect of a 1°C rise in mean global temperature is modest. Predictions of an increase in temperature of about 2–3°C for the 21st century is also not so large on the scale of figure 2. However, considering that this increase comes at a time when the earth is at a maximum mean temperature has serious implications on our sea level and climatic conditions. Even more threatening is the fact that the present global CO_2 level is 380 ppm, which can be seen from figure 2 to be significantly higher than at any time in the past 400,000 years. Getting far out of equilibrium with the known atmospheric ranges that can be tolerated by planet earth is threatening and is having a strong effect on global strategies to increase efforts to greatly lower CO_2 emissions. For example: many European Union nations have pledged to be on 20% renewable energy sources by the year 2020.[5]

Figure 3 shows the implications of the transition to a sustainable energy scenario from today's fossil fuel economy where 85% of today's energy comes from oil, coal and gas with only a few percent coming from renewable sources, such as hydroelectric, wind and photovoltaics. The magnitude of the energy challenge involves both the huge amount of new energy supply that must be found and the short time scale over which this transition to a sustainable energy environment must take place. To achieve such a transition, major advances in basic research, engineering know-how and industrial development will be necessary, along with both energy conservation and better utilization of waste heat for electricity generation, two areas where thermoelectrics can play a major role. To make this happen, higher thermoelectric figure of merit, *ZT,* materials will be needed, as discussed further in this review.

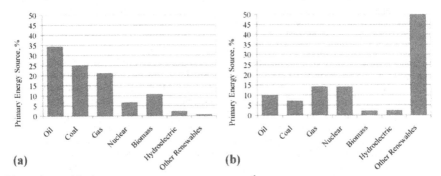

(a)   **(b)**

Figure 3. World primary energy consumption (a) today[6] (2003) and (b) 2050 (target).

THE THERMOELECTRIC MATERIALS CHALLENGE

Nanostructured materials for energy applications

Nanostructured materials are important for energy-based applications overall. When materials have nano-sized features, quantum limits drive the physics, greatly expanding the parameter space for materials properties beyond what can be obtained in conventional materials. Nanostructured materials also can have orders of magnitude more surface area per unit volume, thus promoting catalytic interactions and surface interactions. Since catalysis involves providing energy to overcome an energy barrier, thereby introducing an exponential factor, catalysis allows large benefits to be realized in reaction product for modest increases in thermal energy. Such large scale approaches for change are especially helpful in addressing the daunting scales in magnitude and time of the energy challenge. In addition, engineering materials on the nanoscale can allow independent control of parameters that are interdependent in bulk materials; this aspect of nanomaterials is utilized especially for the case of thermoelectric materials.

An introduction to thermoelectrics

Thermoelectric devices are based on the Seebeck effect and allow for the generation of electrical energy directly from a supply of thermal energy. The Seebeck effect occurs when a material is subjected to a temperature gradient, where electrons diffuse from the hotter region of the material towards the colder region and consequently build up a small voltage – on the order of micro volts per °C temperature difference between the two ends of the material. If dissimilar p- and n-type materials are connected electrically in series and thermally in parallel a more substantial voltage can be developed, which can then be used to supply electrical power to a load. The voltage obtained is directly proportional to the temperature difference that can be maintained: $\Delta V = S\Delta T$. The overall performance depends on three key material properties that affect the device's ability to sustain the temperature difference during operation when current is flowing. Improving thermoelectric efficiency hinges on: increasing the Seebeck coefficient S, so that larger voltages can be obtained for each leg of the thermoelectric device; decreasing the thermal conductivity κ as to minimize the heat being conducted (parasitic heat leakage) through the leg, which warms the cold side and decreases the temperature difference; and increasing the electrical conductivity σ, so that joule heating is minimized (parasitic heat generation), which also serves to warm the cold side.

Thermoelectric materials have been used in a wide variety of applications. Radioisotope generators, which consist of a radioisotope core surrounded by thermoelectric materials coupled to a heat sink, have been used on NASA deep-space and long-life missions to supply power to satillites going to Jupiter and beyond. These thermoelectric devices are reliable and well-proven: the radioisotope generators on many spacecraft, including both Voyager missions launched in 1977, are still operating. Thermoelectrics can be used for earth-based power generation in place of AC generators at remote sites not connected to an electric grid. There is also growing interest in using thermoelectrics to recover waste heat, a potentially huge energy resource. By running the thermoelectric cycle in reverse, thermoelectrics can be used in cooling applications, and are especially effective when traditional working-fluid cooling cycles are not practical, such as miniature refrigerators, computer chip heat sinks, and cooled automobile seats. Thermoelectrics are also used as sensor elements for infrared detectors. The recent explosion in the production of cooled automobile seats has spurred a large expansion in the thermoelectric industry at a time when laboratory demonstrations of enhanced materials performance have been carried out in many laboratories. Therefore the stage is set for an increase in interest in thermoelectric materials containing nanostructures. In this review we discuss examples of major advances that have occurred with the development of SiGe nanocomposite systems and conclude with comments on the potential impact of these research developments on the future directions for the field.

Nanostructured thermoelectric materials

Thermoelectrics are heat engines, limited by the second law of thermodynamics. The efficiency of a thermoelectric generator is related to the dimensionless figure of merit, ZT, which is the product of the temperature, the electrical conductivity, and the square of the Seebeck coefficient, divided by the thermal conductivity: $ZT = T\sigma S^2/\kappa$. In traditional bulk materials, S, σ, and κ are functions of the carrier concentration of the material and cannot be controlled independently. In low-dimensional and other advanced materials, the connection between these parameters can be broken: distorting the density of states can increase the Seebeck coefficient without decreasing the electrical conductivity, and increasing the likelihood of an interface scattering event at a grain boundary that can reduce the thermal conductivity more than the electrical conductivity. Both of these effects have been shown to increase the thermoelectric figure of merit dramatically.

State-of-the-art thermoelectric materials

The current research focus in advancing thermoelectric materials performance is in four areas: phonon glass/electron crystals (PGECs), low-dimensional materials, nanostructured materials, and materials with resonant peaks in the conduction band density of states near the Fermi level, as given in the following examples. PGEC materials such as skutterudites contain atoms which can rattle inside "cages" within the crystal structure; these atoms scatter phonons but not electrons, thus reducing the lattice thermal conductivity without strongly affecting the electrical conductivity. Researchers[7] have measured a ZT of nearly 1.4 at 800 K in the skutterudite $Ba_xYb_yCo_4Sb_{12}$. Low-dimensional materials rely on quantum effects of very small length scales to affect the electronic density of states, and to scatter phonons preferentially to electrons. Two-dimensional Bi_2Te_3/Sb_2Te_3 super-lattices[8] have exhibited a ZT of 2.4 at 300 K. ZT values of 1.6 at 300 K have been achieved in zero-dimensional $PbSe_{0.98}Te_{0.02}/PbTe$ quantum dots[9]. Nanostructured materials, such as BiSbTe that has been ball-milled and hot-pressed, also have exhibited[10] a ZT near 1.4 at about 373 K. Thallium-doped PbTe, a bulk material with a resonant peak in the density of states, has shown[11] a ZT of 1.5 at 773 K. The next phase of the thermoelectrics materials industry seems to require a large expansion of usage. This will require the production of large quantities of thermoelectric modules, and requiring more large-scale production approaches. Here nanocomposite thermoelectric materials are expected to play a special role, once the next generation of materials are better developed.

RECENT ADVANCES IN THERMOELECTRIC MATERIALS SCIENCE AND TECHNOLOGY

Overview of nanocomposites

A promising approach to achieving large quantities of bulk materials with a high ZT is the creation of bulk materials with nano-sized features. One of these approaches is in development of nanocomposite materials for thermoelectric applications. By synthesizing materials with very small grains, phonon scattering can be greatly enhanced, with only a small reduction in electrical conductivity. Production of these materials can be rapidly ramped up because they can be synthesized in large batches.

Bulk materials are the starting components for our nanocomposites. These bulk materials can be based on pure elements, or complex alloys, such as skutterudites. Chunks of these starting materials are mixed with dopants and ball-milled until the mean grain size and grain size distribution is on the appropriate scale for a particular application. For example, ball-milling compounds for a few hours may result in a 20-nm average particle size, but contains a large distribution of subgrains at the 20-nm scale.[10] After the constituents have been mixed and reduced to nanometer size in the ball mill, the resulting powder of nanoparticles is rapidly pressed into ingots with either a DC press or a hot press so that the increase in the grain size distribution is small.

In order to improve ZT, the lattice thermal conductivity must be decreased. Recent modeling has provided insight into the effect of grain orientation and size on phonon transport (see figure 4), and shows that it is the amount of interface area in a sample that scatters phonons more than electrons that dominates the thermal conductivity.[12] It has been shown that in silicon at room temperature, 80% of the lattice thermal conductivity comes from phonons with mean free paths less than 10 μm. Nanostructures restricting the phonon mean free path to 10 nm could reduce the lattice thermal conductivity by nearly two orders of magnitude[13] (as can be seen in figure 5).

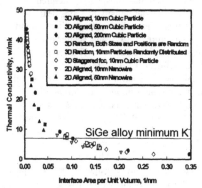

Figure 4. Plot of the thermal conductivity vs. interface area per unit volume shows a universal behavior.[12]

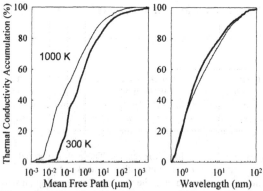

Figure 5. Thermal conductivity accumulation as a function of mean free path (left) and wavelength (right) at two different temperatures.[13]

Modeling of nanocomposite thermoelectrics has shown that there are large temperature ranges over which the power factor can increase and the thermal conductivity can decrease at the same time.[14] Experimental studies of the thermal conductivity and power factor comparing their temperature-dependent performance in 3D bulk materials with our nanocomposite material of the same composition show that these desirable effects only occur in nano-scale materials (see figure

6). Here the power factor and the thermal conductivity for a $Si_{0.80}Ge_{0.20}$ nanocomposite thermoelectric material are compared to the corresponding bulk sample used to power NASA space missions, showing superior thermoelectric performance of nanocomposite $Si_{0.80}Ge_{0.20}$, with characteristics not found in 3D materials.[15] In fact, gains in ZT up to 40% have been made in the Bi_2Te_3 system simply by decreasing the grain size.[10] In these demonstrations of enhanced ZT it is also important to include measurements of devices which exhibit enhanced thermoelectric performance as was done in this case, thereby confirming the findings from the physical properties measurements. Boundaries between nano-sized grains restrict the phonon mean free path through boundary scattering, thereby reducing the lattice contribution to the thermal conductivity. The power factor increase for the nanocomposites largely comes from an increase in carrier density.

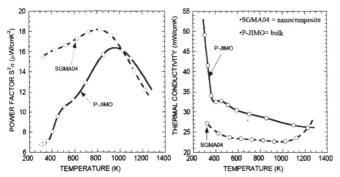

Figure 6. Comparison of thermoelectric properties (power factor on the left and thermal conductivity on the right) for bulk vs. nanostructured materials based on work done at JPL.[15]

RECENT RESULTS ON $Si_{1-x}Ge_x$ NANOCOMPOSITES

In this section we report very promising recent results on the thermoelectric performance of $Si_{1-x}Ge_x$ nanocomposite sytems as an example of a nanocomposite version of a materials system that is now in use by NASA for thermoelectric applications for power generation on space satellites. Bulk $Si_{1-x}Ge_x$ is a good high-temperature thermoelectric material, with a ZT approaching 1 at 1200 K. Both silicon and germanium are commonly used in the semiconductor industry. Silicon is a plentiful material and the grade of material needed for thermoelectric applications is relatively inexpensive and non-toxic. $Si_{1-x}Ge_x$ can easily be doped with donors (n-type) or acceptors (p-type). Both n- and p-type materials benefit from nanostructuring and are stable up to at least 1050 C. Values of $ZT = 1$ for p-type $Si_{0.80}Ge_{0.20}$ and $ZT = 1.3$ for n-type $Si_{0.80}Ge_{0.20}$ nanocomposite materials have been reported[16,17] with substantial enhancement over the performance of their bulk 3D counterparts. The reason for the more than 50% increase in ZT for the p-type $Si_{0.80}Ge_{0.20}$ doped with boron is associated with the increased phonon scattering at the grain boundaries. This arises from the small grains in the powder material produced by the ball milling process, as shown in figure 7, where we see from the x-ray diffraction pattern that the powder is a single-phase material with a mean particle size of 15 nm. TEM images showing particles of a wide range of sizes with many in the ~20 nm size range and most of the particles contain many much smaller multigrains within the large grains. Furthermore, the compaction process to produce the nanocomposite material shows no significant increase in grain size. The very large distribution of grain sizes in figure 7 is highly effective in increasing the scattering of the wide distribution of phonons present in this material, thereby reducing the thermal conductivity by a factor of about 2, as seen in figure 8 where results for $\sigma(T)$, $S(T)$, $\kappa(T)$, and $ZT(T)$ are shown for the nanocomposite material relative to the bulk material with the same $Si_{0.80}Ge_{0.20}$ ratio. By achieving clean boundaries between grains and the full mass density of $Si_{0.80}Ge_{0.20}$ in the nanocomposite material, the high power factor of the bulk material is reproduced in the nanocomposite. The ZT thus achieved with the nanocomposite material is about 90% higher than that of the bulk p-type $Si_{0.80}Ge_{0.20}$ material now flown for NASA missions. The stability of the nanostructure at elevated temperatures (with tests done at 1100° C for 7 days) was investigated by annealing studies at elevated temperatures and the results on the thermoelectric property measurements are shown by the open circles in figure 8.

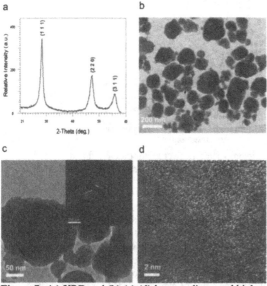

Figure 7. (a) XRD and **(b),(c),(d)** low, medium, and high-resolution TEM characterization studies on p-type $Si_{0.80}Ge_{0.20}$.[16] Inset shows electron diffraction measurements, indicating a collection of crystallites within a single grain.

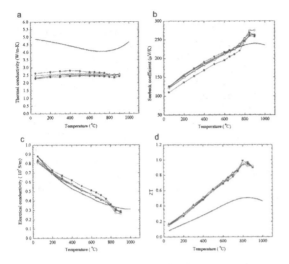

Figure 8. Temperature dependent **(a)** $\kappa(T)$, **(b)** $S(T)$, **(c)** $\sigma(T)$, and **(d)** $ZT(T)$ measurements of p-type nanocomposite $Si_{0.80}Ge_{0.20}$ material compared to 3D bulk material of the same composition (see text).[16]

Similar success was achieved with n-doped $Si_{0.80}Ge_{0.20}$ material doped with phosphorus[17] as shown in figure 9 for which a $ZT \sim 1.3$ was obtained. In this case, a similar scenario to that for the p-type $Si_{0.80}Ge_{0.20}$ nanocomposite material was achieved, with a major reduction in κ while maintaining or slightly increasing the power factor, and maintaining good thermal stability under elevated temperature (1050°C in this case) operation.

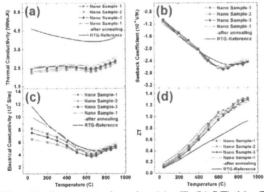

Figure 9. Temperature dependent **(a)** $\kappa(T)$, **(b)** $S(T)$, **(c)** $\sigma(T)$, and **(d)** $ZT(T)$ measurements of n-type nanocomposite $Si_{0.80}Ge_{0.20}$ samples compared to 3D bulk material of the same composition flown by NASA.[17]

37

Since Ge is more expensive and less abundant than Si, there is motivation toward reducing the amount of Ge used in the $Si_{1-x}Ge_x$ alloy. Therefore scientific studies to determine the Ge concentration for the optimization of ZT led to studies of nanocomposites with low Ge concentration, namely $Si_{0.95}Ge_{0.05}$ alloys[18]. Results for the thermoelectric performance shown in figure 10 suggest that 5% Ge is sufficient to make a major reduction in $\kappa(T)$ and to yield an increase in ZT that is comparable to that for the $Si_{0.80}Ge_{0.20}$ bulk thermoelectric material now flown by NASA. Modeling calculations clearly show that alloying strongly reduces the mean free path for phonons with wavelength shorter than ~4 nm due to scattering from point defects and grains of small size. The nanograined material is effective at scattering over the extended range of phonon wavelengths of interest for the $Si_{1-x}Ge_x$ system. The lower Ge concentration,on the other hand, results in a higher carrier mobility. Thus, by varying the doping level and the grain size of nanostructures, some degree of control of the temperature variation of the various thermoelectric parameters can be achieved. This flexibility in materials preparation should prove useful for optimizing the performance of this high temperature thermoelectric system by varying the alloying ratio between Si and Ge.

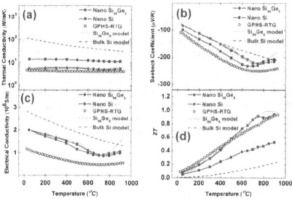

Figure 10. Temperature dependent **(a)** $\kappa(T)$, **(b)** $S(T)$, **(c)** $\sigma(T)$, and **(d)** $ZT(T)$ measurements of bulk nanocomposite $Si_{0.95}Ge_{0.05}$ samples in comparison to those on $Si_{0.80}Ge_{0.20}$ and Si.[18]

OUTLOOK TO THE FUTURE

Capturing waste heat with thermoelectrics

Until now, thermoelectrics have focused on niche applications where their long-term reliability was exploited. With the increased efficiency now coming on line and the increasing global focus on energy conservation and waste heat recovery, we can foresee more widespread use of thermoelectricity in addressing new aspects of the energy challenge. Thermoelectric materials offer tremendous opportunity to convert waste heat into electricity. We can foresee an increasing emphasis given to such uses in the future. As humans burn fuel in automobiles, power plants, manufacturing processes, and buildings, the majority of the fuel's energy is wasted as it passes out tailpipes, up chimneys, or through walls and windows. In each of these situations, a significant amount of the waste heat could be converted by thermoelectrics into useful electricity. Automotive engineers are now working on replacing the alternator in automobiles with thermoelectric devices in the exhaust stream[19]. These thermoelectric devices could intercept 65% of a vehicle's fuel energy now going into the coolant and exhaust streams, thereby resulting in an increase in the fuel efficiency. In residential buildings, thermoelectrics could be implemented in the heating systems to effect a cogeneration system, with the electricity supplementing or ideally replacing the power from the electrical grid. Waste heat from manufacturing processes presently produce an enormous energy stream that has yet to be tapped. Frequently, the waste stream is at a very high temperature, giving high-quality heat that yields a high maximum potential (Carnot) efficiency and should be an attractive target for energy conservation. The increased efficiency of the newly-developed nanocomposite thermoelectric materials described in this review are likely to make such application areas more attractive.

However, as we look into the future, we are even more likely to look for totally new applications of thermoelectric materials that we have not considered before. One such likely direction is a means of capturing solar energy, as discussed below.

Solar power

If we intend to live sustainably on this planet, all our energy must eventually come from renewable and sustainable resources. The sun provides four orders of magnitude more power than humans consume; tapping this resource will be the biggest challenge of the 21st century. Our main tools for harnessing solar energy are photovoltaics, solar thermal systems, and thermoelectrics. The current rate of photovoltaic production and its expected acceleration in the coming years is not at a rate that will allow us to reach a goal of 1.5 TW production by 2050. Photovoltaic cell production would have to follow a trend similar to Moore's Law for semiconductors, with rates doubling every few years if we are to reach expected goals. Thus far, solar thermoelectrics have been largely neglected from widespread consideration, but they could become a key component in a renewable energy portfolio. The promise of solar thermoelectrics is that, unlike photovoltaics, they can use the entire solar spectrum, and they also can recover nearly all the energy of a photon, whereas photovoltaics can only obtain a bandgap of energy from a photon, no matter how energetic it is. Solar thermoelectrics do this more efficiently and at lower cost by focusing the sun's rays onto a selective surface which is black in the solar spectrum and white in the IR, which acts as the hot side for a thermoelectric element. A third technology that draws from both of these concepts is solar thermophotovoltaics, in which the sun's rays are again captured on a selective surface. This thermal energy is then reradiated from the back side of the selective surface onto a parallel photovoltaic cell. The back side of the selective surface is designed to emit at the optimal wavelength for the thermophotovoltaic, maximizing the efficiency of energy conversion from the sun into electricity. Designing these selective absorbers and emitters is challenging, but the payoff is an 85% theoretical maximum efficiency in absorbing the solar energy.

Looking to the Future

Recent advances in thermoelectric materials show promise for a wide range of applications. Laboratory tests have proven new techniques for increasing the thermoelectric figure of merit.

As these results become validated, the thermoelectric field is gaining momentum and is growing. The next step is for researchers and industry to tackle the problems that are now preventing widespread implementation of thermoelectric devices. First, more fundamental studies must be carried out to better understand the physics driving thermoelectric properties. Second, efforts must be made to optimize promising thermoelectric materials systems. Third, industry must attempt to scale up promising materials and approaches in order to make low cost, safe products in large volumes. Finally, there must be appropriate attention given to the interplay between materials properties and system design to maximize total system efficiency. As we continue down this path of increasing efficiency and decreasing cost, new opportunities will present themselves as targets for thermoelectric applications, broadening the scope of the applicability of thermoelectrics in the larger energy landscape.

ACKNOWLEDGMENTS

The authors gratefully acknowledge the assistance of Mario Hofmann for help with the preparation of the figures. MSD, GC, and ZR acknowledge support for this work under DOE Grant DE-FG02-08ER46516. Part of this work was performed at the Jet Propulsion Laboratory, California Institute of Technology under contract with the National Aeronautics and Space Administration.

REFERENCES

1. Population data from 1950 - 2050: Population Division of the Department of Economic and Social Affairs of the United Nations Secretariat. *World Population Prospects: The 2008 Revision.* United Nations, 2008.
2. Population data from 1750 - 1950: United Nations. *The World at Six Billion.* United Nations, 1999.
3. U.S. Department of Energy. Energy Information Administration. *Report# DOE/EIA-0484: International Energy Outlook 2008.* Washington: Government Printing Office (2008).
4. J.T. Houghton *et al., Climate change 2001: The Scientific Basis.* (IPCC, Cambridge, UK, 2001).
5. World International Renewable Energy Conference, Washington, 2008.
6. International Energy Agency. *Energy to 2050: Scenarios for a Sustainable Future.* (IEA, France 2003).
7. X. Shi *et al.,* Appl. Phys. Lett. **92** 182101, (2008).
8. R. Venkatasubramanian, E. Siivola, T. Colpitts, and B. O'Quinn, Nature **413** 597-602 (2001).
9. T.C. Harman, P.J. Taylor, M.P. Walsh, and B.E. LaForge, Science **297** (5590) 2229-2232 (2002).
10. B. Poudel *et al.,* Science **320** (5876) 634-638 (2008).
11. J.P. Heremans *et al.,* Science **321** (5888) 554-557 (2008).
12. R. Yang, Ph.D. dissertation, Massachusetts Institute of Technology, 2006.
13. A.S. Henry and G. Chen, J. Comput. Theor. Nanosci., **5**, 141-152 (2008).
14. A.J. Minnich, M.S. Dresselhaus, Z.F. Ren, and G.Chen, Energy Environ. Sci. (in press).
15. M.S. Dresselhaus *et al.,* Advanced Materials, **19**, 1043-1053 (2007).
16. G. Joshi *et al.,* Nano Lett. **8** (12) 4670-4674 (2008).
17. X.W. Wang *et al.,* Appl. Phys. Lett. **93** 193121 (2008).

18. G. Zhu *et al.*, Phys. Rev. Lett. (in press).
19. K. Matsubara, in *21st International Conference on Thermoelectrics*, Portland, OR, 2002, pp. 418-423.

Mater. Res. Soc. Symp. Proc. Vol. 1166 © 2009 Materials Research Society

Interdiffusion of Bi and Sb in Superlattices Built From Blocks of Bi₂Te₃, Sb₂Te₃ and TiTe₂

Clay D. Mortensen and David C. Johnson
Department of Chemistry and Materials Science Institute, 1253 University of Oregon,
Eugene, Oregon 97403, U.S.A.

ABSTRACT

The reaction kinetics of $[(Ti-Te)]_x[(Sb-Te)]_y$, $[(Bi-Te)]_x[(Sb-Te)]_y$, $[(Ti-Te)]_w[(Bi-Te)]_x$ and $[(Ti-Te)]_w[(Bi-Te)]_x[(Ti-Te)]_y[(Sb-Te)]_z$ precursors as a function of annealing temperature and time was probed using x-ray diffraction techniques to define the parameters required to form superlattice structures. $[(TiTe_2)_{1.36}]_x[Sb_2Te_3]_y$ and $[(TiTe_2)_{1.36}]_x[Bi_2Te_3]_y$ superlattices were observed to form while $[(Bi-Te)]_x[(Sb-Te)]_y$ precursors yielded only $Bi_{2-x}Sb_xTe_3$ alloys. This behavior was correlated with the miscibility/immiscibility of the constituents of the targeted superlattices. For the three component system, Bi and Sb were observed to interdiffuse through the Ti-Te layer over the range of Ti-Te thicknesses explored, resulting in formation of $(Bi_xSb_{1-x})_2Te_3$ alloys within the superlattice structure. When the Bi_2Te_3 and Sb_2Te_3 thicknesses were equal, symmetric $[\{(TiTe_2)\}_{1.36}]_w[(Bi_{0.5}Sb_{0.5})_2Te_3]_y$ superlattices were formed.

INTRODUCTION

Over the past 30 years, nanostructured $(A)_x(B)_y$ superlattices have allowed researchers to tune physical properties including magnetism[1] and thermal conductivity[2] and design new semiconductor structures with enhanced performance for applications.[3] Increasing the complexity by adding a third component in an $(A)_x(B)_y(C)_z$ pattern has resulted in exciting breakthroughs including increased polarization enhancement in ferroelectric materials,[4,5] delta doping of semiconductors,[6] high mobility semiconductor heterostructures,[7] and tuning of optoelectronic properties[8] not achievable with $(A)_x(B)_y$ superlattices.

Three component superlattice structures have been prepared with a variety of epitaxial growth techniques (atomic layer deposition, pulsed laser deposition, chemical vapor deposition, and molecular beam epitaxy) to sequentially deposit the three superlattice components, but it is frequently difficult to find suitable growth conditions. Molecular beam epitaxy is generally limited by the need for epitaxial relationships and limited and often conflicting growth conditions. Low temperatures are desired to limit losses due to the high vapor pressure of some elements and to limit interdiffusion of the elements. High temperatures are necessary to obtain sufficient surface mobility. As the number of components increase, the lack of favorable growth conditions and epitaxial relationships limits the material combinations that can be explored for the discovery of unprecedented novel properties.[9,10]

Recently Harris and coworkers reported a new solid-phase growth technique that used self assembly of preconfigured reactants to prepare a series of $[(Bi_2Te_3)]_x[(TiTe_2)_{1.36}]_y$ materials having a 36% lattice mismatch between the a-b planes of the components.[11,12] In this paper, we examine the reaction kinetics of $[(Ti-Te)]_x[(Sb-Te)]_y$, $[(Bi-Te)]_x[(Sb-Te)]_y$, $[(Ti-Te)]_w[(Bi-Te)]_x$ and $[(Ti-Te)]_w[(Bi-Te)]_x[(Ti-Te)]_y[(Sb-Te)]_z$ precursors as a function of annealing temperature and time using x-ray diffraction techniques to define the parameters required to form superlattice structures. Over the range of Ti-Te thicknesses used, Bi and Sb interdiffuse through the Ti-Te layer resulting in formation of $(Bi_xSb_{1-x})_2Te_3$ layers. When the Bi_2Te_3 and Sb_2Te_3 thicknesses were equal, $[\{(TiTe_2)\}_{1.35}]_w[(Bi_{0.5}Sb_{0.5})_2Te_3]_y$ superlattices are formed.

EXPERIMENT

Bismuth, titanium, antimony and tellurium (99.995% purity) were obtained from Alfa Aesar and used without further purification. All thin films of this study were deposited at a background pressure of <6 x 10^{-7} Torr onto polished silicon substrates. Veeco Applied Epi SUMO effusion cells were used to deposit Bi, Sb and Te. Heating the temperature of the tip and the base of the effusion cell with separate PID controllers controlled the deposition rate. Inficon quartz crystal microbalances were used to monitor the deposition rate and they were located 10 inches above each elemental source. Titanium was deposited using a 6 kV Thermionics Laboratory electron beam gun. A constant rate of 1 Å/s was maintained using power control feedback via the quartz crystal microbalance. Six-inch silicon substrates were positioned 30 inches above the sources and the substrates were rotated during the deposition to improve the thickness uniformity. Pneumatic source shutters controlled via a custom LabView interface were used to control the thickness and sequence of the elemental layers to prepare the targeted superlattice precursors. Actual deposition thicknesses were determined experimentally with x-ray reflectivity.

The precursors were subjected to annealing in a nitrogen environment (<0.5 ppm O_2) within a box furnace to promote self-assembly of the multilayer precursors into the desired crystalline product.

X-ray reflectivity (XRR) and High Angle X-ray Diffraction (HAXRD) measurements were performed using a Bruker D8 Discover Diffractometer (Cu Kα radiation, λ = 1.5418 Å). A θ-2θ configuration was used for XRR, with 2θ scanned from 0 to 10° with a step size of 0.01° and a counting time of 1 sec per step. A θ-2θ configuration was also used for HAXRD, with 2θ scanned from 5 or 10° to 65° using a step size of 0.05° and a counting time of 3 sec per step. Rocking curve measurements obtained as a function of annealing temperature and time were used to assess the degree to which the a-b planes of the superlattice were aligned with the silicon substrate. Synchrotron x-ray radiation experiments were performed at the Advanced Photon Source at Argonne National Laboratory using beam lines 33 BM-C and 33-ID-D. In-plane diffraction scans were obtained by setting the sample at a small incident angle (0.20 degrees) and scanning the detector in the a-b plane of the sample.

A Cameca SX-50 or SX-100 Electron microprobe was used to obtain the composition of the films discussed in this paper. For each sample, 10 data points were collected on each sample and the results were averaged together to obtain the composition of the films. The films on a silicon substrate were analyzed using three different accelerating voltages (8, 12, and 16 kV), and the StrataGem software package was used to determine the composition of the films. Raw intensities were corrected using procedures developed by Donovan and Tingle.[13] StrataGem performs quantitative chemical analysis by comparing experimental k-ratios to simulated values, and employs the XPP formalism developed by Pouchou and Pichoir.[14,15]

DISCUSSION

Precursor design and calibration of deposition parameters

The initial step in the synthesis procedure developed by Harris is to determine the deposition conditions required to deposit a bilayer of the constituent elements that has the correct composition and contains the quantity of atoms required to form a single layer of the desired Te-Ti-Te, Te-Bi-Te-Bi-Te or Te-Sb-Te-Sb-Te structures. This calibration procedure accounts for the different sticking coefficients of the elements, the difference between the monitored vs. actual thickness of the deposited layers, and the unknown densities of the amorphous elemental layers. A total of 14 samples were deposited to calibrate the ratio of deposition times required to obtain the correct composition for the three targeted constituents; Sb_2Te_3, $TiTe_2$ and Bi_2Te_3. A linear regression of the inverse atomic percent of metal (Bi, Sb or Ti respectively) versus the ratio of deposition times of Te to metal (Bi, Sb or Ti respectively) yielded the ratio of deposition times required to obtain the desired stoichiometry. During initial annealing studies of precursors designed to yield the binary compounds, a 3-5 % contraction of the film was observed. Presumably this is due to an increase in packing efficiency of the elemental layers on compound formation. Also, a slight excess of tellurium (1-3%) was found to increase crystal growth and compensated for a similar percentage of tellurium lost during annealing.

To determine the deposition times of Bi and Te required to form one 10.16 Å Te-Bi-Te-Bi-Te layer of Bi_2Te_3, the deposition times of Sb and Te required to form one 10.12 Å Te-Sb-Te-Sb-Te layer of Sb_2Te_3 and the deposition times of Ti and Te required to form one 6.49 Å Te-Ti-Te layer of $TiTe_2$, a series of multilayer precursors were prepared. To calibrate the Bi-Te repeat thickness, samples were deposited holding the number of Ti-Te and Sb-Te layers constant at 3 while systematically varying the number of Bi-Te repeats from 2-6, resulting in five multilayers with increasing multilayer period thickness. The resulting systematic change in the multilayer repeat thickness was used to determine the thickness of one Bi-Te layer by linear regression. While holding the ratio of deposition times constant to maintain the stoichiometry of each building block, the deposition times of Bi and Te were adjusted to yield a systematic increase in the multilayer period by 10.1 Å for each Bi-Te. Using the same iterative approach, repeat unit thicknesses obtained for one binary reaction couple of Sb-Te and Ti-Te resulted in systematic changes in the multilayer period by 10.1 Å and 6.5 Å respectively.

Using these calibrated deposition times, a series of samples were deposited for the intended study and the thickness and composition of the films were measured with XRR and EPMA respectively. The thicknesses agreed with the expected sums of the individual layers and the measured compositions were compared to the expected composition based upon the formula unit $[(TiTe_2)_{1.36}]_x[(Bi_2Te_3)]_y-[(TiTe_2)_{1.36}]_x[(Sb_2Te_3)]_z$,where the subscript 1.36 accounts for the lattice mismatch of $TiTe_2$ relative to Bi_2Te_3 and Sb_2Te_3. The change in the atomic composition ratios of Bi:Sb, Bi:Ti, Bi:Te, Sb:Ti, Sb:Te and Ti:Te were used for comparison as the ratios are much more sensitive to small changes in the composition. The error from the EPMA composition determination was propagated through the ratio determination at the 95% confidence interval. As shown in Figure 1, the composition changes regularly with each change in the number of binary reactants, and the experimentally achieved compositions are within error of the expected composition. The thickness and composition data verifies that the precursor films contain the desired nanoarchitecture and compositions to form the targeted superlattice structures.

45

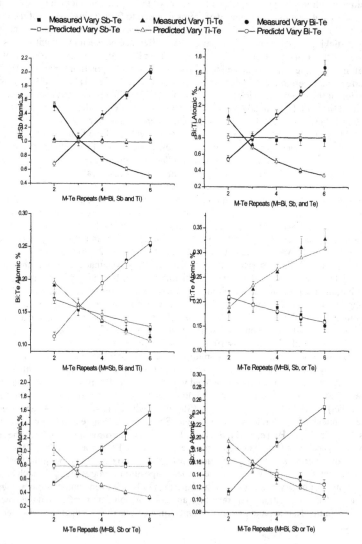

Figure 1. Atomic ratios of Bi:Sb, Bi:Ti, Sb:Ti, Sb:Te, Ti:Te and Bi:Te of the films prepared during the absolute layer thickness calibration.

Kinetics of Crystallization

To explore the kinetics of the competing effects of interdiffusion, phase separation and immiscibility on superlattice formation, $[(Bi-Te)]_x[(Sb-Te)]_y$, $[(Ti-Te)]_x[(Sb-Te)]_y$ and $[(Bi-Te)]_x[(Ti-Te)]_y$ precursors were prepared from combinations of Bi-Te, Sb-Te and Ti-Te binary reaction couples and XRR and HAXRD collected as a function of temperature and time.

$[(Bi-Te)]_x[(Sb-Te)]_y$ precursors

The data collected on the $[(Bi-Te)]_3[(Sb-Te)]_3$ precursor, prepared by depositing 3 alternating Bi-Te layers followed by 3 alternating Sb-Te layers, is representative of this family of precursors. In the XRR data, shown in Figure 2, only the first order of Bragg diffraction is observed, corresponding to a calculated multilayer thickness of 63 Å. Simulations predict that 3 orders of Bragg diffraction maxima should be observed below 10 degrees 2-theta, suggesting interdiffusion of Bi and Sb during deposition. Annealing results in a decrease in the absolute intensity of the Bragg peak from the as-deposited layering with complete loss of Bragg diffraction above 150°C. Concurrently, the (0 0 3) reflection of $(Bi_xSb_{1-x})_2Te_3$ at 7.8 2-theta grows as annealing temperature is increased. In the HAXRD data, also shown in Figure 2, diffraction maxima expected for $(Bi_xSb_{1-x})_2Te_3$ grow in intensity as annealing temperature is increased. No diffraction maxima corresponding to an extended superstructure are observed, leading to the conclusion that the initially compositionally modulated layers mix to form $(Bi_{0.5}Sb_{0.5})_2Te_3$. In-plane diffraction scans obtained at the Advanced Photon Source contained only reflections for the $(Bi_{0.5}Sb_{0.5})_2Te_3$ alloy at all stages of annealing. This data suggests that the targeted superlattice $[(Bi_2Te_3)]_3[(Sb_2Te_3)]_3$ does not occur because interdiffusion of Bi and Sb occurs faster than crystal growth of the superlattice.

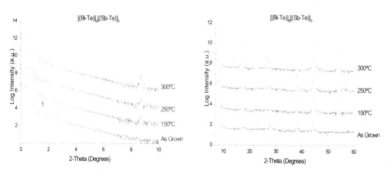

Figure 2. Evolution of the $[(Bi-Te)]_3[(Sb-Te)]_3$ precursor as a function of annealing temperature.

47

[(Bi-Te)]ₓ[(Ti-Te)]ᵧ precursors

The data collected on the [(Bi-Te)]₃[(Ti-Te)]₃ precursor, prepared by depositing 3 alternating Bi-Te layers followed by 3 alternating Ti-Te layers, is representative of this family of precursors. Three orders of Bragg diffraction were observed in the XRR data (Figure 3) of the as deposited precursor, corresponding to a 51.8 Å superlattice period. On annealing the precursor at higher temperature, two additional orders of superlattice diffraction maxima are observed in the XRR data and the superlattice period decreases to 50.8 Å. The intensity of the Bragg peaks diminished after annealing at 300°C, suggesting interdiffusion or phase separation of the superlattice. In the HAXRD data, broad diffraction maxima are observed corresponding to superstructure formation in the as deposited sample. Higher temperature annealing of the precursor results in the appearance and growth of 35 orders of Bragg diffraction maxima that can be indexed as the (0 0 L) family of reflections for a [(Bi₂Te₃)]₃[(TiTe₂)₁.₃₆]₃ superlattice. The most intense superlattice diffraction pattern occurs at 250°C, and higher temperature annealing results in phase separation of the superlattice. The immiscibility of Bi₂Te₃ and TiTe₂ do not provide a driving force for mixing, allowing the formation of the superlattice. Higher annealing temperatures increase the diffusion rate, leading to the phase separation of the superlattice into its respective components.

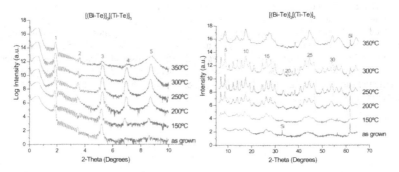

Figure 3. Evolution of the [(Bi-Te)]₃[(Ti-Te)]₃ precursor as a function of annealing temperature.

[(Sb-Te)]ₓ[(Ti-Te)]ᵧ precursors

The data collected on the [(Sb-Te)]₃[(Ti-Te)]₃ precursor, prepared by depositing 3 alternating Sb-Te layers followed by 3 alternating Ti-Te layers, is representative of this family of precursors. Four orders of Bragg diffraction maxima are observed in the XRR data of the as deposited sample, corresponding to a superlattice period of 54.6 Å. As shown in Figure 4, the intensity of the Bragg peaks increases with annealing temperature up to 300° and the superlattice period decreases slightly to 50.7 Å. Higher annealing temperatures result in a decrease in the absolute intensity of the Bragg peaks. In the HAXRD data, over 30 orders of Bragg diffraction maxima are observed as the sample is annealed as a [(Sb₂Te₃)]₃[(TiTe₂)₁.₃₆]₃ superlattice forms. Higher annealing temperatures results in a decrease in Bragg peak intensity and reflections from the binary constituents can be observed, suggesting that the superlattice phase separates into the binary components. There is no driving force for mixing because Sb_2Te_3 and $TiTe_2$ are immiscible.

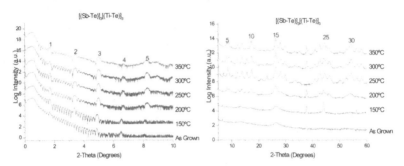

Figure 4. Evolution of the $[(Sb\text{-}Te)]_3[(Ti\text{-}Te)]_3$ precursor as a function of annealing temperature.

$[(Bi\text{-}Te)]_z[(Ti\text{-}Te)]_y[(Sb\text{-}Te)]_x[(Ti\text{-}Te)]_y$ precursors

Based upon the results of the initial annealing studies of the two component precursors, there are two possible pathways for $[(Bi\text{-}Te)]_z[(Ti\text{-}Te)]_y[(Sb\text{-}Te)]_x[(Ti\text{-}Te)]_y$ precursors to evolve with increasing annealing temperature. The precursor could self assembly into a $[(Bi_2Te_3)]_z[(TiTe_2)_{1.36}]_y[(Sb_2Te_3)]_x[(TiTe_2)_{1.36}]_y$ superlattice if the Ti-Te layer acts as a diffusion barrier, stopping Bi and Sb from mixing. Alternatively, the Bi and Sb layers could mix through the Ti-Te layer resulting in a superlattice containing alternating layers of $TiTe_2$ and a $(Bi_xSb_{1-x})_2Te_3$ alloy.

To differentiate between these two possibilities, a $[(Ti\text{-}Te)]_3[(Bi\text{-}Te)]_3[(Ti\text{-}Te)]_3[(Sb\text{-}Te)]_3$ precursor was prepared and diffraction data collected as a function of annealing temperature and time. If the Bi and Sb layers do not mix, a $[(Bi_2Te_3)]_3[(TiTe_2)_{1.36}]_3[(Sb_2Te_3)]_3[(TiTe_2)_{1.36}]_3$ superlattice with a period of 101.6Å is expected to form. If the Bi and Sb layers mix, interdiffusion results in the formation of a $[(TiTe_2)_{1.36}]_3[((Sb_{0.5}Bi_{0.5})_2Te_3)]_3$ superlattice with a period of 50.8 Å because the Bi_2Te_3 and Sb_2Te_3 thicknesses are equal.

Figure 5 contains the XRR and HAXRD data collected as a function of annealing temperature to probe the reaction pathway of the $[(Ti\text{-}Te)]_3[(Bi\text{-}Te)]_3[(Ti\text{-}Te)]_3[(Sb\text{-}Te)]_3$ precursor. As deposited the first 13 orders of Bragg diffraction are observed corresponding to a superlattice period of 103.7 Å. On annealing at higher temperatures the absolute intensity of the Bragg peaks increases as long range order of the superlattice improves. Above 200 °C, every odd order Bragg peak begins to decreased in intensity until they can no longer be observed. The resulting diffraction pattern can be indexed with a unit cell half the size of the initial period. To confirm that the decrease in intensity occurs as a result of interdiffusion of Bi and Sb, in-plane x-ray diffraction scan was obtained on the precursor after annealing at 250°C. All of the observed diffraction maxima could be indexed to the (h k 0) family of diffraction maxima of

49

$(Bi_{0.5}Sb_{0.5})_2Te_3$ and $TiTe_2$ and the calculated a-lattice parameters of $TiTe_2$ (3.77 Å) and $(Bi_{0.5}Sb_{0.5})_2Te_3$ (3.28 Å) agree very well with previously published lattice parameters for the compounds. Increasing the thickness of the Ti-Te layer up to 40 Å did not prevent the interdiffusion of the bismuth and antimony.

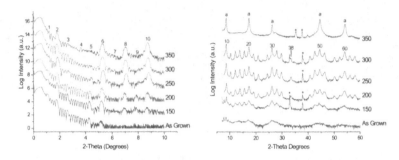

Figure 5. XRR and HAXRD showing the evolution of the $[(Ti-Te)]_3[(Bi-Te)]_3[(Ti-Te)]_3[(Sb-Te)]_3$ precursor with annealing.

CONCLUSIONS

This work demonstrate that to prepare $(A)_x(B)_y(C)_z$ superlattice structures from modulated elemental reactants, all three components must be immiscible in one another to reduce the enthalpy of mixing. To form $(A)_x(B)_y(A)_x(C)_y$ superlattices all three components must either be immiscible or the layer A must be an effective interdiffusion barrier to keep B and C from mixing.

ACKNOWLEDGMENTS

This work was supported by the Office of Naval Research (N0014-07-1-0358). Clay Mortensen acknowledges an IGERT fellowship supported by the National Science Foundation (DGE-0549503). The use of the APS was supported by the U.S. Department of Energy, Office of Science, Office of Basic Energy Sciences, under Contract No. DE-AC02-06CH11357. The assistance of Dr. Paul Zschack in obtaining and analyzing the in-plane diffraction data is gratefully acknowledged.

REFERENCES

1. J. Shen, J. Kirschner, Surf. Sci. **500**, 300 (2002).
2. D. G. Cahill, W. K. Ford, K. E. Goodson, G. D. Mahan, A. Majumdar, H. J. Maris, R. Merlin, S. R. Phillpot, Journal of Applied Physics **93 (2)**, 793 (2003).
3. R. Venkatasubramanian, E. Siivola, T. Colpitts, B. O'Quinn, Nature **413**, 597 (2001).
4. H. N. Lee, H. M. Christen, M. F. Chisholm, C. M. Rouleau, D. H. Lowndes, Nature **433**, 395 (2005).
5   S. M. Nakhmanson, K. M. Rabe, D. Vanderbilt, Appl. Phys. Lett. **87**, 102906 (2005).

6. S. M. Landi, C. V. Tribuzy, P. L. Souza, R. Butendeich, A. C. Bittencourt, G. E. Marques, Phys. Rev. B: Condensed Matter and Materials Physics **67 (8)**, 085304/1-085304/10 (2003).
7. P. Hashemi, L. Gomez, J. L. Hoyt, M. D. Robertson, M. Canonico, Appl. Phys. Lett. **91 (8)**, 083109/1-083109/3 (2007).
8. L. Seravalli, P. Frigeri, M. Minelli, P. Allegri, V. Avanzini, S. Franchi, Appl. Phys. Lett. **87 (6)**, 063101/1-063101/3 (2005).
9. A. G. Norman, T. Y. Seong, I. T. Ferfuson, G. R. Booker, B. A. Joyce, Semicond. Sci. technol. **8**, S9 (1993).
10. M. Takahasi, J. Mizuki, J. Cryst. Growth **275**, 2201 (2005).
11. F. R. Harris, S. Standridge, D. C. Johnson, J Am Chem Soc **127**, 7843 (2005).
12. F. R. Harris, S. Standridge, C. Feik, D. C. Johnson, Angewandte Chemie Int. Ed. **42(43)**, 5296 (2003).
13. J. J. Donovan, T. N. Tingle, J. Microscopy **2**, 1 (1996).
14. J. L. Pouchou and F. Pichoir, Scanning **12**, 212 (1990).
15. J. L. Pouchou, F. Pichoir, and D. Boivin, Microbeam Analysis, San Francisco Press: San Francisco, 120 (1990).

51

Mater. Res. Soc. Symp. Proc. Vol. 1166 © 2009 Materials Research Society

High Temperature Thermoelectric Properties of Nano-Bulk Silicon and Silicon Germanium

Sabah Bux[1], Jean-Pierre Fleurial[2], Richard Blair[3], Pawan Gogna[2], Thierry Caillat[2], and Richard Kaner[1]

[1]Department of Chemistry and California NanoSystems Institute, University of California, Los Angeles, Los Angeles, California;
[2]Power and Sensor Systems Group, Jet Propulsion Laboratory/California Institute of Technology, Pasadena, California;
[3]Department of Chemistry, University of Central Florida, Orlando, Florida.

ABSTRACT

Point defect scattering via the formation of solid solutions to reduce the lattice thermal conductivity has been an effective method for increasing ZT in state-of-the-art thermoelectric materials such as Si-Ge, Bi_2Te_3-Sb_2Te_3 and PbTe-SnTe. However, increases in ZT are limited by a concurrent decrease in charge carrier mobility values. The search for effective methods for decoupling electronic and thermal transport led to the study of low dimensional thin film and wire structures, in particular because scattering rates for phonons and electrons can be better independently controlled. While promising results have been achieved on several material systems, integration of low dimensional structures into practical power generation devices that need to operate across large temperature differential is extremely challenging. We present achieving similar effects on the bulk scale via high pressure sintering of doped Si and Si-Ge nanoparticles. The nanoparticles are prepared via high energy ball milling of the pure elements. The nanostructure of the materials is confirmed by powder X-ray diffraction and transmission electron microscopy. Thermal conductivity measurements on the densified pellets show a drastic reduction in the lattice contribution at room temperature when compared to doped single crystal Si. The combination of low thermal conductivity and high power factor leads to an unprecedented increase in ZT at 1275 K by a factor of 3.5 in n-type nanobulk Si over that of single crystalline samples. Experimental results on both n-type and p-type Si are discussed in terms of the impact of the size distribution of the nanoparticles, doping impurities and nanoparticle synthesis processes.

INTRODUCTION

With the recent revelations on the impact of fossil fuels on the environment thermoelectric energy conversion as a possible cleaner power source has experienced strong interest. Prior to the renewed interest in thermoelectrics, the National Aeronautics and Space Administration (NASA) has had a successful history of utilizing thermoelectrics to power deep space science probes and planetary missions for the past forty years [1]. Although the current design for the radioisotope thermoelectric generator has been both successful and reliable, NASA anticipates that in order for future missions to be effective, significant improvements in the conversion efficiency and the specific power (W/kg) must be made. Despite a great deal of research on thermoelectric properties over the past 30 years, the maximum ZT value of currently available bulk materials is still limited to about 1.2 over a temperature range of 100-1500 K [2].

Decoupling phonon scattering and charge carrier scattering mechanisms to maximize ZT has been a key strategy. Two leading approaches have been examined in order to separately optimize electronic and thermal transport by either investigating complex structures such as Skutterudites and Zintl phases or by developing low dimensional structures. It was theorized that a large increase in the figure of merit could be achieved by increasing the density of states through quantum confinement of charge carriers [3, 4]. Soon after the publication of the theoretical paper, promising experimental results were reported on thin film Bi_2Te_3 superlattices [5], PbTe thick film quantum dot superlattices [5] and more recently Si nanowires [6, 7]. Detailed analyses of these results indicated that the most significant contribution to the enhancement in ZT was due to a large reduction in lattice thermal conductivity (κ_l). Despite the significant increases in ZT, such low dimensional structures are ill suited for high temperature power generation applications due to the need for stable operation at temperatures up to 1275 K and device integration with heat sources. Additionally, processing and integration of such materials would be difficult and costly to scale-up for applications such as radioisotope thermoelectric generators and large scale high grade waste heat recovery.

It has been demonstrated that producing nanoscale features on the bulk scale can lead to significant enhancements in ZT, mostly due to the reduction in the lattice thermal conductivity via random interfacial scattering of the reduced phonon mean free path. Most of this work has focused on low to intermediate temperature materials such as Bi_2Te_3 [9], $CoSb_3$ [10], and PbTe [11]. In order to address materials that can operate at high temperature, we report on an approach to synthesizing bulk scale nanostructured Si and $Si_{1-x}Ge_x$ semiconductors.

EXPERIMENT

The unfunctionalized n-type doped nanosilicon ($Si_{0.9640}P_{0.0216}Ga P_{0.0072}$), n-type nanosilicon-germanium ($Si_{1-x}Ge_xP_{0.020}$), p-type nanosilicon ($Si_{0.980}B_{0.020}$) and p-type nanosilicon-germanium ($Si_{1-x}Ge_xB_{0.020}$) were prepared by mechanical alloying of high purity 325 mesh Si powder and or Ge lump, with lump P and sometimes GaP pieces or 20 mesh B powder. Stoichiometric amounts of the respective powders were processed in inert environments using tungsten carbide vials loaded into a customized dual mixer mill. After several hours of mixing, the powders were isolated and then hot pressed using high density graphite dies at a temperature in excess of 1275 K and a pressure of 150 MPa under argon. The hot pressed pellets were approximately 12.7 mm x 15 mm. The pellets were then sliced using a diamond saw into 1 mm x 12.7 mm wafers for characterization and property measurements. The geometric density of the pellets was determined to approach 99% of the theoretical density.

The thermoelectric transport properties were measured on the full size hot pressed pellets as well as thin slices cut from each pellet. Transport property measurements were conducted of as a function of temperature up to 1275 K using both commercial and custom-made setups that have been described elsewhere [12-14]. Prior to any measurement the samples were "reset" to the same initial high doping level by a rapid thermal anneal at 1325 K followed by an air quench to preserve the high temperature solubility of the dopant in the Si lattice. The properties measured included electrical resistivity, Hall coefficient and Seebeck coefficient as well as thermal diffusivity all of which were measured using a heating rate of 180 K/hr. Thermal conductivity was measured using a Netzch laser flash diffusivity system from room temperature to 1275 K. The lattice thermal conductivity was calculated by subtracting the electronic contribution of the carriers from the total thermal conductivity term $\kappa_{lattice} = \kappa_{total} - \kappa_{electronic}$. The electronic thermal conductivity was calculated by the Wiedermann-Franz law $\kappa_{electronic} = L\sigma T$ where L is the Lorenz number and σ is the electrical conductivity. Since our samples are heavily doped, a value of 2.2×10^{-8} $J^2.K^{-2}.C^{-2}$ was used for the Lorenz number [15].

DISCUSSION
Synthesis and Characterization of Bulk Scale Nanostructure Materials

In order to synthesize large (10-15g) quantities of unfunctionalized silicon and silicon-germanium nanoparticles smaller than 50 nm in size, the technique of high energy ball milling was used. It is an effective, straightforward and industrially scalable technique of synthesizing thermoelectric materials and materials with unfavorable phase diagrams such as Si-Ge [16, 17]. Recently, experimental work as a function of the ball milling parameters has been conducted on the state-of-the-art $Si_{0.8}Ge_{0.2}$ alloy composition and Si and some of these results have been reported recently [18-22]. The nanostructured materials are made via a series of repeated impact collisions in an agitated ball mill. The nanostructure of the resulting products was confirmed using several techniques. Since the nanoparticles are unfunctionalized, they tend to form large aggregates and therefore a combination of characterization methods was used. Powder X-ray diffraction of nanostuctured Si produced by this method (Figure 1a) shows that the product is phase pure and free of vial contamination. Line broadening analysis using the method of integral breadths was used to calculate the average crystallite size of 15 nm. A TEM image of a typical Si nanoparticle aggregate is shown in Figure 1b. The aggregate size ranges from 50 to 100 nm. However, higher resolution imaging demonstrates that the aggregate is actually made up of 15-20 nm crystallites, which is in agreement with the value calculated from XRD (Figure 1c). Upon high temperature compaction, TEM imaging on an ion milled sample illustrates that 15 nm nanoscale domains remain after high temperature/high pressure compaction (Figure 1d).

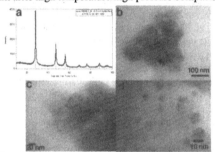

Figure 1. a) Powder X-Ray diffraction pattern of nanobulk Si, the lines correspond to JCPDS reference pattern for Si. **b)** TEM image of nanobulk Si aggregate. **c)** Higher resolution image of image b, showing nanoscale features of the nanobulk aggregate. **d)** TEM image of ion milled densified pellet. Nanoscale features are preserved after high temperature processing.

Thermoelectric Properties

Recently, we have reported that bulk scale n-type nanostructured Si (nanobulk Si) exhibits up to 90% reduction in the lattice thermal conductivity when compared to single crystal Si (Figure 2a) [22]. We attribute this mainly to increased phonon scattering by the nanocrystalline domains. The lattice thermal conductivity was reduced from as high as 870 $mW.cm^{-1}.K^{-1}$ for single crystal to 62 $mW.cm^{-1}.K^{-1}$ for nanobulk Si at 300 K, reaching a minimum of 30 $mW.cm^{-1}.K^{-1}$ at 1200 K. Concurrently, the carrier mobility was only reduced by 30% for the Si samples. The significant reduction in the lattice thermal conductivity and minimal degradation in the carrier mobility for n-type nanobulk Si leads to an unprecedented 250% increase in the ZT of Si (Figure 2b).

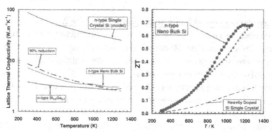

Figure 2. a) Lattice thermal conductivity of n-type nanobulk Si. The lattice thermal conductivity of n-type nanobulk Si approaches that of the heritage $Si_{0.80}Ge_{0.20}$ alloy. b) ZT of n-type nanobulk Si. Data for only two samples are shown for clarity.

When the process of producing nanobulk Si via high energy ball milling is applied to p-type Si, the results are remarkably different. Measurements on compacted p-type nanobulk Si do not exhibit as significant a decrease in thermal conductivity, only being reduced by 60% when compared to the 90% reduction in n-type nanobulk Si. In addition, the carrier mobility suffers a greater setback by being reduced by nearly 50%. However, when a small amount of Ge (2.5 % atomic) is added to the p-type nanobulk Si system, the lattice thermal conductivity decreases dramatically from 305 $mW.cm^{-1}.K^{-1}$ for p-type nanobulk Si to a value of 70 $mW.cm^{-1}.K^{-1}$ for 2.5 % Ge alloyed nanobulk Si at 300 K. In order to evaluate the magnitude of electron-phonon transport decoupling and its impact on ZT, we calculated values for the parameter, β, which is proportional to $\mu_o(m^*)^{3/2}/\kappa_L$, where μ_o is the carrier mobility, m^* is the carrier effective mass [23]. From Fermi-Dirac statistics, peak ZT values at the optimum doping level increase with increasing β values. Assuming minimal change in the carrier effective mass for dilute Si-rich n- and p-type $Si_{1-x}Ge_x$ alloys, we examine the experimental μ_H/κ_L ratios at 300 K in figure 3a. There is a significant impact to the μ_H/κ_L with increasing Ge content and nanostructuring for samples at similar carrier concentrations. It is interesting to note that the p-type samples have a much sharper increase in the μ_H/κ_L going from Si to diluted Si-Ge alloys compared to n-type samples. However, nanostructuring alone is more effective in pure phosphorus doped n-type Si even though boron doping in p-type should act as a stronger point defect scattering center.

Figure 3b represents the experimental maximum ZT (ZT_{max}) at 1275 K for nanostructured samples with varying Ge concentrations. Figure 3b mimics the trends seen in figure 3a; with increasing Ge content, the maximum ZT at 1275 K increases for both n- and p-type $Si_{1-x}Ge_x$ demonstrating enhancements over the heritage Si-Ge technology. Once again, the slope for the increase in ZT for p-type $Si_{1-x}Ge_x$ is much steeper than for n-type. At this time it is unclear as to why the carrier mobilities of both n- and p-type systems are reduced. At such high carrier concentrations the electron mean free path in the samples should only be a few nanometers, suggesting that grain boundary scattering of the carriers should not be very effective. Potential barriers at the grain boundaries, intra-grain scattering due to nanocrystalline domains, or the formation of a low mobility impurity band are all possible mechanisms [24]. We are in the process of conducting extensive studies on the electronic effects of point defect scattering and examining the mobility at different carrier concentrations.

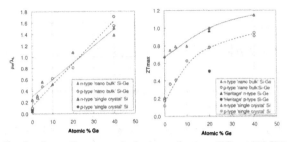

Figure 3 a). Ratio of μ_H/λ_L vs Ge content for both n and p-type SiGe at 300 K b) ZTmax at 1275 K for n- and p-type SiGe with increasing Ge content.

The thermal stability of the nanobulk Si samples was examined by varying several processing conditions which ranged from compacting the samples at high temperature (1475 K) to longer anneal times at temperatures in excess of 1275 K. Upon completion, the samples did not exhibit noticeable grain growth, and all had similar thermal conductivity values and similar thermoelectric properties, thus demonstrating the stability of the nanostructures. Similarly, long term "life time" tests, over 4500 hours at 1275 K, on nanostructured $Si_{0.80}Ge_{0.20}$ samples for high temperature generators do not show changes in thermal conductivity values (Figure 4). Due to the more refractory nature of Si, it is expected that nanobulk Si will behave in a similar manner.

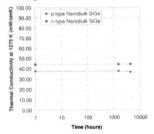

Figure 4. Thermal stability of nanostructured of n- and p-type $Si_{0.80}Ge_{0.20}$.

CONCLUSIONS

We have demonstrated that n-type nanobulk Si is a much more effective thermoelectric material with a ZT enhancement of over 250% over single crystal. This is mainly attributed to a significant decrease in the lattice thermal conductivity and minimal degradation in the carrier mobility. The combination of point defect scattering and nanostructuring also leads to significant ZT enhancements in both p-type and n-type SiGe. Nanobulk Si (n-type) with "dilute" Si rich $Si_{1-x}Ge_x$ for p-type is demonstrated as a potentially promising and cost effective material combination for large scale waste heat recovery.

ACKNOWLEDGMENTS

The authors would like to thank support from the National Science Foundation DMR 0805352 (RBK), an IGERT Fellowship DGE-0114443 and DGE-0654431 (SKB). Part of this work was performed at the Jet Propulsion Laboratory, California Institute of Technology under

REFERENCES

1. D. M. Rowe (Ed) *CRC Handbook of Thermoelectrics*, CRC, Boca Raton, Florida, USA (1995).
2. G. J. Synder, E. S. Toberer. *Nat. Mater.* **7**, 105 (2008).
3. L. D. Hicks and M. S. Dresselhaus, *Physical Review B*, **47**, 16631 (1993).
4. L. D. Hicks and M. S. Dresselhaus, *Physical Review B*, **47**, 12727 (1993)
5. R. Venkatasubramanian, E. Siivola, T. Colpitts, B. O'Quinn. *Nature.* **413**, 597 (2001).
6. T. C. Harman, P. J. Taylor, M. P. Walsh, B. E. LaForge. *Science.* **297**, 2229 (2002).
7. A. Boukai, Y. Bunimovich, T.-K. Jamil, J.-K. Yu, W. A. Goddard, J. R. Heath. *Nature.* **451**, 168 (2008)
8. A. I. Hochbaum, R. Chen, R. D. Delgado, W. Liang, E. C. Garnett, M. Najarian, A. Majumdar, P. Yang. *Nature.* **451**,163 (2008).
9. X. B. Zhao, T. J. Zhu, X. H. Ji. in *CRC Handbook of Thermoelectrics Macro to Nano.* (Ed. D. M. Rowe), Boca-Raton, Florida, USA, 24-1 (2006).
10. X. Ji, J. He, P. Alboni, Z. Su, N. Gothard, B. Zhang, T. M. Tritt, J. W. Kolis. *Phys. Stat. Sol. (RRL).* **1**, 229 (2007).
11. K. F. Hsu, S. Loo, F. Guo, W. Chen, J. S. Dyck, C. Uher, T. Hogan, E. K. Polychroniadis, M. G. Kanatzidis. *Science.* **303**, 818 (2004).
12. J. W. Vandersande, C. Wood, A. Zoltan, D. Whittenberger. *Thermal Conductivity,* **19**, 445 (1988).
13. J. A. McCormack, J.-P. Fleurial. *Materials Research Society Symposium Proceedings.* **234**, 135 (1991).
14. C. Wood, D. Zoltan, G. Stapfer. *Review of Scientific Instruments.* **56**, 719 (1985).
15. V. I. Fistul. *Heavily Doped Semiconductors.* (Translated by A. Tybulewicz), New York, New York, USA, (1969).
16. B. A. Cook, J. L. Harringa, S. H. Han in *CRC Handbook of Thermoelectrics.* (Ed. D. M. Rowe), Boca-Raton, Florida, USA, 124 (1995).
17. B. A. Cook, B. J. Beaudry, J. L. Harringa, W. J. Barnett. *Proceedings of the International Energy Conversion Engineering Conference.* 693 (1989).
18. M. S. Dresselhaus, G. Chen, M. Y. Tang, R. G. Yang, H. Lee, D. Z. Wang, Z. F. Ren, J.-P. Fleurial, P. Gogna. *Adv. Mater.* **19**, 1043 (2007).
19. M. S. Dresselhaus, G. Chen, Z. Ren, J.-P. Fleurial, P. Gogna, M. Y. Tang, D. Vashaee, H. Lee, X. Wang, G. Joshi, G. Zhu, D. Wang, R. Blair, S. Bux, R. Kaner. *Proceedings of the 2007 Fall Materials Research Society Meeting on Thermoelectrics.* 1044-U02-04 (2007).
20. G. Joshi, H. Lee, Y. Lan, X. Wang, G. Zhu, D. Wang, R. W. Gould, D. C. Cuff, M. Y. Tang, M. S. Dresselhaus, G. Chen, Z. Ren. *Nano Lett.* **8**, 12, 4670 (2008).
21. X. W. Wang, H. Lee, Y. C. Lan, G. H. Zhu, G. Joshi, D. Z. Wang, J. Yang, A. J. Muto, M. Y. Tang, J. Klatsy, S. Song, M. S. Dresselhaus, G. Chen, Z. F. Ren. *App. Phys. Lett.* **93**, 193121 (2008).
22. S. Bux, R. Blair, P. Gogna, H. Lee, G. Chen, M. Dresselhaus, R. Kaner, J.-P. Fleurial. *Adv. Funct. Mater.* **19**, 1 (2009).
23. C. B. Vining. *J. Appl. Phys.* **69**, 331, (1991).
24. D. L. Young, H. M. Branz, F. Liu, R. Reedy, B. To, Q. Wang. *J. Appl. Phys.* **105**, 033715 (2009).

Mater. Res. Soc. Symp. Proc. Vol. 1166 © 2009 Materials Research Society

Investigation of Solid-State Immiscibility and Thermoelectric Properties of the System PbTe – PbS

Steven N. Girard,[1] Jiaqing He,[2] Vinayak P. Dravid,[2] and Mercouri G. Kanatzidis[1]
[1]Department of Chemistry, Northwestern University, Evanston IL 60208
[2]Department of Materials Science and Engineering, Northwestern University, Evanston IL 60208

ABSTRACT

We have shown that $(Pb_{1-m}Sn_mTe)_{1-x}(PbS)_x$ where m = 0.05 and x = 0.08 exhibits a ZT of ~1.4 at 700 K. This system incorporates two thermoelectric systems: PbS_xTe_{1-x} and $Pb_{1-x}Sn_xTe$. Here we report the thermoelectric properties of PbS_xTe_{1-x} (x = 0.08 and 0.30). The material $PbS_{0.08}Te_{0.92}$ exhibits nucleation and growth of PbS precipitates, while $PbS_{0.30}Te_{0.70}$ exhibits PbS precipitation through spinodal decomposition phase separation. We report the thermoelectric properties of this system as a result of the differing precipitation phenomena.

INTRODUCTION

Modern initiatives to develop renewable energy resources have propelled thermoelectric heat-to-power generation as an attractive option. The efficiency of a thermoelectric material is related to the figure of merit ZT, defined as $ZT = S^2\sigma T/\kappa$, where S is the thermopower or Seebeck coefficient, σ is the electrical conductivity, T is the operating temperature, and κ is the total thermal conductivity (a sum of the electronic κ_{elec} and lattice κ_{lat} vibrations). Many conventional thermoelectric materials have limited applications as a result of only a moderate ZT of approximately 1. By embedding nanostructured phases into a conventional material, the lattice thermal conductivity can be significantly reduced without altering electronic transport, resulting in an enhanced ZT. This has been demonstrated on a number of thin-film[1] and bulk systems.[2-6]

The material $(Pb_{0.95}Sn_{0.05}Te)_{0.92}(PbS)_{0.08}$ was shown to be an excellent thermoelectric material, reaching a ZT of ~1.4 at 650 K.[7] This system was designed to incorporate two bulk thermoelectric materials: PbS_xTe_{1-x} and $Pb_{1-x}Sn_xTe$. $Pb_{1-x}Sn_xTe$ is a robust thermoelectric alloy exhibiting solid solution behavior with a beneficial reduction in κ_{lat} at low temperature due to alloy point defect scattering of phonons. In addition, this material also can exhibit p-type conduction at large enough Sn concentrations.[8] Conversely, the material PbS_xTe_{1-x} has been shown to produce a two-phase PbTe – PbS nanocomposite; toward the middle of the composition range PbS will precipitate from the matrix by spinodal decomposition resulting in band-like structures, whereas toward the outer composition range PbS will precipitate by nucleation and growth resulting in distinct particles.[9] This report focuses mainly on phase immiscibility within the PbTe – PbS system. The PbTe – SnTe system has been thoroughly studied previously.[8, 10, 11]

EXPERIMENT

Samples of PbS_xTe_{1-x} were synthesized using PbTe and PbS starting materials prepared using high-purity starting elements by conventional high temperature solid-state method. Starting materials were flame sealed in fused silica ampoules at a residual vacuum of ~10^{-4} Torr and heated to 1050° C in a box furnace. Samples were inverted several times at the melt and allowed to cool with the furnace door open to allow rapid solidification. The resulting ingots were free of bubbles and cracks and could be easily cut and polished for properties measurements. Purity of phases and homogeneity along the ingot was confirmed by powder X-ray diffraction (PXRD).

DISCUSSION

According to the phase diagram for the PbTe – PbS system, rapid solidification to room temperature will result in the local structure being completely homogeneous, i.e. a solid solution.[12] Any thermal treatment within the chemical spinodal region or miscibility gap will result in precipitation of PbS. We found that following rapid solidification of PbS_xTe_{1-x} samples, a solid solution alloy was created, as evidenced by high resolution transmission electron microscopy (HRTEM) and selected area electron diffraction (SAED), Figure 1a. The precipitation of PbS was observed in-situ in electron-transparent foils of $PbS_{0.30}Te_{0.70}$ mounted on a heating stage and heated to 475 K, Figure 1b.

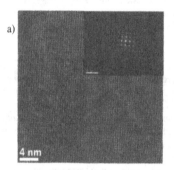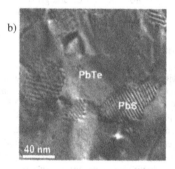

Figure 1. a) HRTEM image of $PbS_{0.30}Te_{0.70}$ rapidly cooled, showing a single-phase material, inset shows single phase by SAED, b) TEM image of $PbS_{0.30}Te_{0.70}$ heated in-situ mounted on a heating stage at 475 K shows precipitation of PbS.

The creation of a solid solution and the precipitation of PbS in $PbS_{0.30}Te_{0.70}$ was also confirmed by powder X-ray diffraction (PXRD), Figure 2.

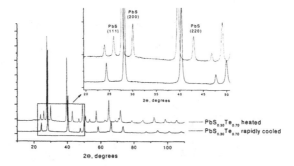

Figure 2. PXRD patterns of rapidly cooled and heated $PbS_{0.30}Te_{0.70}$. The technique of rapidly cooling creates solid solution alloys of PbS_xTe_{1-x}, and a post-anneal will cause PbS precipitation from the PbTe matrix.

The microstructure of finely-polished samples (post-heating) were observed using an FEI Helios Nanolab dual ion beam (FIB)/scanning electron microscope (SEM) following a cleaning etch using the ion beam. Preferential etching of the minor phases was observed and allowed a more clear view of the microstructure with minimal surface deformation. Samples of $PbS_{0.08}Te_{0.92}$ revealed mostly spherical particles of PbS homogeneously embedded throughout the PbTe matrix, Figure 3a. The average size of the particles identified by SEM appeared ~50 nm in diameter. Samples of $PbS_{0.30}Te_{0.70}$ revealed PbS bands 50-100 nm in width that extended microns in length, Figure 3b. These observations are in complete agreement with the classical model of nucleation and growth and spinodal decomposition phenomena.

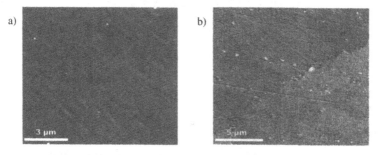

Figure 3. Microstructure of a) $PbS_{0.08}Te_{0.92}$, revealing particles of PbS (dark spots) ~50 nm in diameter formed through a nucleation and growth process, b) $PbS_{0.30}Te_{0.70}$, revealing a labrythine structure of PbS bands (dark areas) 50-100 nm in width that extend many microns in length formed through a spinodal decomposition process. Sample surfaces were etched using a Ga^+ ion beam, resulting preferential etching of the PbS phase.

High resolution images of the nanostructure of these samples were observed using a JEOL-2100 HRTEM. For $PbS_{0.08}Te_{0.92}$, smaller particles can be observed within the PbTe as small as 5 nm in diameter, Figure 4a. Within the larger particles, there appeared to be local compositional inhomogeneity, potentially resulting in nanostructures within the PbS precipitates, Figure 4b. The same is observed within the PbS bands in $PbS_{0.30}Te_{0.70}$. In both cases, the regions of PbS were revealed to be fully coherent within the PbTe matrix as shown by SAED, Figure 4c inset.

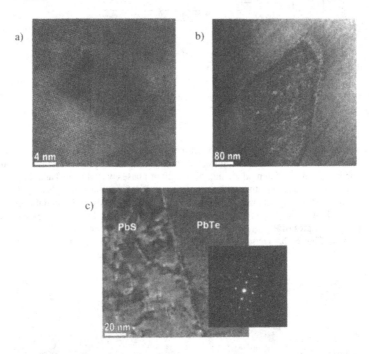

Figure 4. HRTEM images of a) detail of small precipitate in $PbS_{0.08}Te_{0.92}$, b) large PbS precipitate, containing smaller particles and compositional inhomogeneities, c) detail of interface between PbS band and PbTe matrix in $PbS_{0.30}Te_{0.70}$. The PbS precipitates are fully coherent within the PbTe matrix, as shown by SAED (inset).

Electrical transport measurements were completed using an ULVAC ZEM-3 electrical conductivity and Seebeck coefficient system. The precipitation of PbS from the PbTe matrix could be observed in rapidly cooled samples, as evidenced by a marked change in electrical transport measurements.

Notable is the effect of a solid solution alloy of $PbS_{0.08}Te_{0.92}$ versus $PbS_{0.30}Te_{0.70}$ and their respective composite systems on electrical transport. For $PbS_{0.08}Te_{0.92}$, the minor PbS phase is metastable, requiring a sufficient amount of thermal energy to overcome an activation energy barrier to allow for PbS precipitation.[12] It would appear that the barrier was overcome at around 400 K, resulting in a deviation from the normal power law scaling expected for a PbTe-based material. The resulting electrical conductivity increased while thermopower decreased, suggesting an increase in carrier concentration and electron mobility.

For $PbS_{0.30}Te_{0.70}$, the material is unstable, resulting in no activation energy barrier to nucleation of a PbS phase.[12] As a result, electrical transport changed as a function of temperature throughout the measurement. In contrast to $PbS_{0.08}Te_{0.92}$, the resulting composite $PbS_{0.30}Te_{0.70}$ had reduced electrical conductivity and increased thermopower, perhaps the result of decreased carrier concentration caused by PbS precipitation.

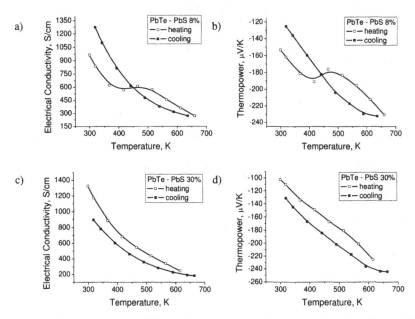

Figure 5. Electrical transport of PbTe – PbS samples; a) electrical conductivity of PbTe – PbS 8%, b) thermopower of PbTe – PbS 8%, c) electrical conductivity of PbTe – PbS 30%, d) thermopower of PbTe – PbS 30%.

CONCLUSIONS

We showed that by rapidly cooling PbS_xTe_{1-x}, a solid solution alloy can be created and thermoelectric properties measured. The precipitation of PbS was observed in-situ as the material was heated to 700 K. It appeared that PbS was precipitated in the material $PbS_{0.08}Te_{0.92}$ around 400 K, and that precipitation of PbS in $PbS_{0.30}Te_{0.70}$ occurred immediately upon heating. The electrical transport for both systems changed significantly depending on composition: for $PbS_{0.08}Te_{0.92}$, an increase in electrical conductivity and a decrease in themopower suggested an increase in carrier concentration and mobility, while the opposite may be true for $PbS_{0.30}Te_{0.70}$.

ACKNOWLEDGEMENTS

The authors thank the Office of Naval Research for funding.

REFERENCES

1.   H. Böttner, G. Chen, and R. Venkatasubramanian, *MRS Bull.* **31**, 211 (2006).
2.   G. S. Nolas, J. Poon, and M. G. Kanatzidis, *MRS Bull.* **31**, 199 (2006).
3.   K. F. Hsu, S. Loo, F. Guo, W. Chen, J. S. Dyck, C. Uher, T. Hogan, E. K. Polychroniadis, and M. G. Kanatzidis, *Science* **303**, 818 (2004).
4.   P. F. P. Poudeu, J. D'Angelo, A. D. Downey, J. L. Short, T. P. Hogan, and M. G. Kanatzidis, *Angew. Chem. Int. Ed.* **45**, 3835 (2006).
5.   J. Androulakis; K. F. Hsu, R. Pcionek, H. Kong, C. Uher, J. J. D'Angelo, A. Downey, T. Hogan, and M. G. Kanatzidis, *Adv. Mat.* **18**, 1170 (2006).
6.   J. R. Sootsman, R. J. Pcionek, H. Kong, C. Uher, and M. G. Kanatzidis, *Chem. Mater.* **18**, 4993 (2006).
7.   J. Androulakis, C. -H. Lin, H. -J. Kong, C. Uher, C. -I. Wu, T. Hogan, B. A. Cook, T. Caillat, K. M. Paraskevopoulos, and M. G. Kanatzidis, *J. Am. Chem. Soc.* **129**, 9780 (2007).
8.   M. Orihashi, Y. Noda, L. D. Chen, T. Goto, and T. Hirai, *J. Phys. Chem. Solids* **61**, 919 (2000).
9.   M. S. Darrow, W. B. White, and R. Roy, *Mater. Sci. Eng.* **3**, 289 (1969).
10. F. Wald, *Energy Convers.* **8**, 135 (1968).
11. Y. Gelbstein, G. Gotesman, Y. Lishzinker, Z. Dashevsky, and M. P. Dariel, *Scr. Mater.* **58**, 251 (2008).
12. J. D. Gunton and M. Droz, *Lecture Notes in Physics: Introduction to the Theory of Metastable and Unstable States*, Vol. 183 (Springer-Verlag, Berlin, Heidelberg, New York, Tokyo, 1983) pp. 1-13.

Mater. Res. Soc. Symp. Proc. Vol. 1166 © 2009 Materials Research Society　　　1166-N03-12

IR Reflectivity Studies of Mechanically Alloyed PbTe Nanocrystals

Thomas Ch. Hasapis[1], Chrysi Papageorgiou[2], Euripides Hatzikraniotis[1], Theodora Kyratsi[2] and Konstantinos M Paraskevopoulos[1]

[1]Department of Physics, Aristotle University of Thessaloniki, 54123 Thessaloniki, Greece
[2]Department of Mechanical and Manufacturing Engineering, University of Cyprus, 1678 Nicosia, Cyprus

ABSTRACT

Nano-crystalline lead telluride powder was synthesized by mechanical alloying using a high-energy planetary ball mill. The broadening of the X-ray diffraction peaks vs ball milling time, indicates small crystalline size of the order of 30nm. IR spectroscopy results are discussed and compared to the material prepared from melt.

INTRODUCTION

Nanostructuring is one of the effective approaches to increase the figure-of- merit of thermoelectric materials. Recent work has shown that small, dimensionally-confined systems can exhibit figures of merit well in excess of one [1]. A major trend in recent research involves the incorporation of nanoscale constituents within bulk materials to form nanocomposites [2]. Additionally, application of low cost techniques on the preparation of nano-materials is of great interest. Research has shown the importance of nano-structured materials and the effect of the nanofeatures on the enhancement of the thermoelectric performance. However, due to the difficulty in applying expensive fabrication techniques in commercial systems, a high-ZT-low-cost approach on nanomaterials is of great interest. Therefore, techniques such as ball milling and sintering are being investigated because of their advantages, which include, beyond the cost, easy shaping and mass production.

In this work, nano-crystalline lead telluride powder was synthesized by mechanical alloying using a high energy planetary ball milling system. Phase transformation and crystallite size evolution during ball milling was followed by powder X-ray diffraction (PXRD) and the morphology was studied by scanning electron microscopy (SEM). IR spectroscopy results are discussed and compared to the material prepared from melt.

EXPERIMENTAL

PbTe was synthesized by mechanical alloying in a Fritsch planetary ball mill, using pure 99.999% Pb and Te elements, as starting materials, taking the processing route described in ref [3]. The speed was 400 rpm and the initial ball-to-material ratio 10:1. Ball milling was interrupted several times (at 1h, 3h, 6h, 12h and 31h) and powder was taken from the batch to be examined by PXRD (Shimadzu), SEM (TESCAN) and TEM (JEOL) system. After the end of ball milling at 31h, the powder was sintered at $700^{o}C$ for 48 hours, and this sample was taken as a reference for PXRD analysis. Infrared spectra (113V Bruker spectrophotometer) were carried out on cold pressed pellets using about 0.7 GPa pressure. The reflection coefficient was determined by typical sample-in, sample-out method with a gold mirror as the reference. Infrared measurements were taken at near normal incidence, with a resolution of $2cm^{-1}$. IR spectroscopy results are compared to those from the material prepared from melt; poly-crystalline material was first grounded using mortar and pestle and then cold pressed at the same pressure.

RESULTS and DISCUSSION

Structural and morphology studies

The XRD patterns for different ball milling times are shown in Fig. 1a. As shown, the produced phase is pure PbTe, at least after 6h of ball milling, which is in agreement with the findings in ref. 3. As is evident, there is a broadening of the peaks compared to the reference (annealed) sample. From the broadening of the peaks, using the well-known Scherrer method [4], the average observed crystallite size is calculated to be of the order of 30nm.

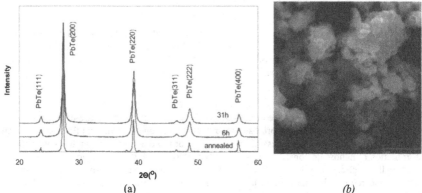

(a)   (b)

Figure 1: (a) XRD pattern evolution for different ball-milling times (b) SEM micrograph for material after 31hrs ball milling.

In order to further study the structural changes, SEM images were taken as shown in Fig. 1b. Based on these images, the typical size of the particles in ball-milled samples is found to be <2μm. TEM diffraction images for the annealed samples present a crystalline cubic phase, while bright field images show large crystallites in different orientations. In ball-milled samples diffractions present, mainly, rings with spots and bright field images show clusters of nano-crystalline areas against a background of a different relief, while further studies are in progress.

IR Reflectivity studies

The IR spectrum (shown in Fig.2) for PbTe is well known for years. PbTe belongs to a class of semiconductor materials that exhibit significant dispersion of optical phonon modes. Pure PbTe is characterized by a TO phonon at $\sim32cm^{-1}$ and the corresponding LO at $\sim104cm^{-1}$ [5]. In Fig. 2a are presented the spectra of PbTe samples synthesized by ball milling at different times of activation. These spectra are analyzed in order the spectroscopic parameters to be received.

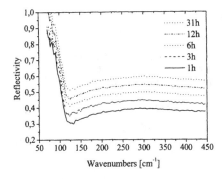
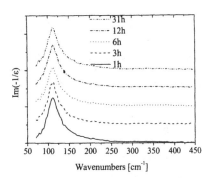

Figure 2: (a) IR reflectivity spectra and (b) the corresponding Im[-1/ε] for different ball milled PbTe samples. Spectra are shifted vertically for clarity

The IR reflectivity (R) is given by the expression:

$$R(\omega) = \left| \frac{\sqrt{\varepsilon(\omega)} - 1}{\sqrt{\varepsilon(\omega)} + 1} \right|^2 \qquad (1)$$

where, $\varepsilon(\omega)$ is the complex dielectric function ($\varepsilon = \varepsilon_1 + i\cdot\varepsilon_2$), which, in the case of PbTe, is expressed in terms of the TO and LO frequencies (ω_T and ω_L) and the plasma frequency (ω_P) as:

$$\varepsilon(\omega) = \varepsilon_\infty \prod_j \frac{\omega_{L,j}^2 - \omega^2 + i\cdot\gamma_{L,j}\omega}{\omega_{T,j}^2 - \omega^2 + i\cdot\gamma_{T,j}\omega} - \frac{\varepsilon_\infty \omega_P^2}{\omega^2 + i\cdot\gamma_P\omega} \qquad (2)$$

where γ_T, γ_L and γ_P are the damping constants for the phonons and the plasmon, and ε_∞ is the high frequency value for the dielectric function. Analysis results are summarized in Table I.

Table I: IR analysis results for sample taken from melt and sintered sample

	From melt		Sintered		Plasmon		
	j=1	j=2	j=1	j=2		from melt	Sintered
ω_L (cm^{-1})	101.4	82.9	100.3	81.0	ω_P (cm^{-1})	56.0	61.8
γ_L (cm^{-1})	23.6	30.6	16.8	30.8	γ_P (cm^{-1})	26.9	36.4
ω_T (cm^{-1})	32.0	79.7	32.0	80.6	N (cm^{-3})	$0.76\cdot10^{17}$	$0.97\cdot10^{17}$
γ_T(cm^{-1})	10.0	17.5	10.0	15.8	ε_∞	21.8	22.7

As can be seen in Table I, two phonons are required to fit reflectivity spectra for samples taken from melt as well as for the sintered samples. The 1st phonon (ω_T=32cm^{-1}, ω_L~100cm^{-1}) is the typical phonon for PbTe, while the 2nd one (ω_T=80cm^{-1}), results from the edge of the Brillouin zone [6]. Plasma frequency is related to the free carrier concentration (N) by the expression:

$$\omega_P^2 = \frac{N \cdot e^2}{m^* \varepsilon_0 \varepsilon_\infty} \qquad (3)$$

where $m^*=0.1m_e$ is the free carrier effective mass [7], ε_0 the permittivity of vacuum and e the free electron charge. Typically, in PbTe, plasmon is coupled with the $\omega_T=32\text{cm}^{-1}$, $\omega_L\sim100\text{cm}^{-1}$ phonon, and the two coupled modes appear in the Im($-1/\varepsilon$) curve (Fig. 2b), at frequencies [8]:

$$2 \cdot \omega_\pm^2 = \left(\omega_L^2 + \omega_P^2\right) \pm \sqrt{\left(\omega_L^2 - \omega_P^2\right)^2 + 4\omega_P^2\omega_T^2} \qquad (4)$$

The graph of eq. 4 is presented in Fig. 3b. As can be seen, due to the low value of plasma frequency, the ω^+ mode is phonon like, and appears as a strong peak in the Im($-1/\varepsilon$), while the ω^- mode, is not shown, as our data begin from 70cm^{-1}.

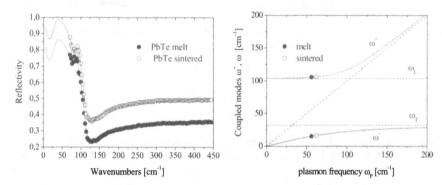

Figure 3: (a) IR reflectivity spectra for PbTe from melt and sintered, along with the calculated spectra by dispersion analysis. Spectra are shifted vertically for clarity. (b) The dependence of ω^+ and ω^- optical coupled modes as a function of plasma frequency (ω_P).

Application of the Effective medium theory in the analysis of IR Reflectivity

Since ball milled samples are grounded and then cold pressed, a significant surface roughness is expected, and this manifests itself by a lowering trend in the reflectivity at high frequencies. Assuming a Gaussian distribution of surface roughness, the reflectivity of the rough surface (R_{SR}) is related to the reflectivity of a smooth one (R_0) by [9]

$$R_{SR} \approx R_0 \cdot \exp\left[-\left(\frac{4\pi \cdot \sigma \cdot n_0}{\lambda_0}\right)^2\right] \qquad (5)$$

where λ_0 is the vacuum wavelength and σ is the RMS value of the surface roughness. The effect of surface roughness, corrected, is presented in Fig.4a.

In Table II the values of density of the cold pressed pellets, as a function of the ball milling time are given. Density was calculated by the mass/volume ratio. As can be seen, there is a decrease of the observed density from 7.6 to 6.6 g/cm^3. We performed an effective medium analysis, taking the Bruggerman approximation [10], and three phases, namely, the microcrystalline phase, with the characteristics taken from sintered sample, a B phase (nano-crystalline clusters) and the air inclusions. The 3-phase Bruggerman approximation for the dielectric function (ε_{BR}) is given as:

$$\sum_j f_j \frac{\varepsilon_j - \varepsilon_{BR}}{\varepsilon_j + 2\varepsilon_{BR}} = 0 \qquad (6)$$

where ε_j is the dielectric function of each of the three phases considered, and f_j their volume fraction. The volume fraction f_3 (the air inclusions) was taken constant (f_3= 0.066) for all samples, as estimated from the sample prepared from melt and since pellet fabrication conditions were the same. The other two (f_1 and f_2) were calculated from density and the condition Σf_j=1. Calculated values for f_j are presented, also, in Table II.

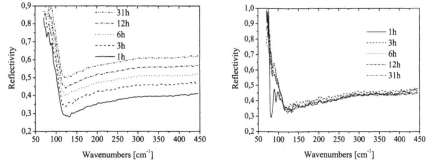

Figure 4: (a) IR reflectivity spectra for ball milled PbTe of Fig. 2(a), after the surface roughness correction. Spectra are shifted vertically for clarity.
(b) Reconstructed IR reflectivity spectra for the phase "B".

In Figure 4b, the reconstructed IR spectra for the phase "B" are shown. As can be seen, apart from the 1h ball milled sample, which clearly deviates from the rest, presumably due to incomplete reaction, all the rest follow the same pattern. This is a clear indication for the co-existence of phase "B" which is present even from the 3h ball milling time, and its volume fraction tends to increase, as ball milling time progresses.

Table II: Volume fractions of different phases and IR results for various ball-milled samples

Sample	d (g/cm³)	Relative d (%)	f_1	f_2	ω_P (cm⁻¹)	γ_P (cm⁻¹)	ε_∞	N (cm⁻³)
from melt	7.61	92.6%	0.934	-	56.0	26.9	21.8	$0.76 \cdot 10^{17}$
1h	6.92	84.1%	0.692	0.242	52.5	77.4	21.2	$0.65 \cdot 10^{17}$
3h	7.03	85.5%	0.582	0.352	49.9	78.2	22.9	$0.64 \cdot 10^{17}$
6h	6.57	79.9%	0.513	0.421	52.3	77.7	22.9	$0.7 \cdot 10^{17}$
12h	6.78	82.5%	0.446	0.488	52.7	78.4	22.8	$0.7 \cdot 10^{17}$
31h	6.61	80.4%	0.356	0.578	52.0	74.2	22.8	$0.69 \cdot 10^{17}$

CONCLUSIONS

In this work, nano-crystalline lead telluride powder was synthesized by mechanical alloying using a planetary ball mill. Phase transformation and crystallite size evolution during ball milling was followed by powder X-ray diffraction (PXRD) and the morphology was studied by scanning electron microscopy (SEM). From the broadening of the PXRD peaks in ball milled samples an average crystalline size in the order of 30nm was obtained. TEM shows clusters of nano-crystalline areas against a background of a different relief. IR spectroscopy results indicate the presence of this nano-crystalline phase and analysis shows that the volume fraction tends to increase, as ball milling time progresses. From the values of optical parameters (ω_P and γ_P) it is evident that ball milling of Pb and Te powders produces samples with low free carrier concentration ($\omega_P{\sim}52\text{cm}^{-1}$ that corresponds to the order of 10^{17}cm^{-3}) which is typical for undopped PbTe [11] and with relatively large damping factor ($\gamma_P{\sim}78\text{cm}^{-1}$). These values are slightly improved in sintered samples ($\omega_P{\sim}62\text{cm}^{-1}$ and $\gamma_P{\sim}36\text{cm}^{-1}$).

ACKNOWLEDGMENTS

The Greek authors would like to acknowledge financial support from European Community and the General Secretariat for Research and Technology-Hellas in the framework of the program PENED 2003 (03EΔ887) and the Hellenic Telecommunications Organization (OTE S.A.). The UCY authors acknowledge financial support from INTERREG IIIA Greece-Cyprus and Cyprus Promotion Foundation (PENEK/ENISX/ 0308/43).

REFERENCES

1. *Thermoelectrics Handbook: Macro to Nano*, (2006) Ed: D. M. Rowe, CRC Press
2. M.S. Dresselhaus, G. Chen, M.Y. Tang, R. Yang, H. Lee, D.Z. Wang, Z.F. Ren, J.P. Fleurial, and P. Gogna, (2007) *New Directions for Low-Dimensional Thermoelectric Materials*, Advanced Materials, 19, 1043-1053,
3. N. Bouad, R.M. Marin-Ayral, G. Nabiasa, J.C. Tedenac, (2002). *Phase transformation study of Pb–Te powders during mechanical alloying*, J. All.Comp. 353, 184-188
4. J.I Langford, A.J.C Wilson, (1978) *Scherrer after 60 years: A Survey and some new Results in the Determination of Crystalline Size*, J. Appl. Cryst. 11, 102 -112
5. J.R. Dixon and H.R. Riedl , (1965) *Electric-Susceptibility Hole Mass of Lead Telluride* Phys. Rev. 138, A873- A881
6. H. R. Riedl, J. R. Dixon and R. B. Schoolar, (1967) *Reflectivity of Tin Telluride in the Infrared* Phys. Rev. 162, 692-700
7. H.R. Riedl, *Free-Carrier Absorption in p-type PbTe*, Phys. Rev., 127, 162 (1962)
8. A. A. Kukharskii, (1973) *Plasmon-Phonon Coupling in GaAs* Solid State Communications, 13, 1761-1765
9. P. Beckman and A. Spizzichino, (1963) *The scattering of electromagnetic waves from rough surfaces*, Pergamon Press, Oxford
10. D.A.G. Bruggeman, (1935) *Berechnung verschiedener physikalischer Konstantenvon heterogenen Substanzen*, Ann. Phys. (Leipzig) 24, 636-679
11. (a) Landolt-Börnstein, III/41: Semiconductors –Subvolume III/41C: Non-tetrahedrally Bonded Elements and Binary Compounds I, Springer Series in Solid State Sciences (b) Schlicht, B., Dornhaus, K., Nimtz, G., Haas, L. D., Jakobus, T.: Solid State Electron. 21 (1978) 1481.

Mater. Res. Soc. Symp. Proc. Vol. 1166 © 2009 Materials Research Society 1166-N03-27

Enhanced Thermoelectric Properties in PbTe Nanocomposites

H. Kirby,[1] J. Martin,[1] A. Datta,[1] L. Chen,[2] and G.S. Nolas[1]

[1]Department of Physics, University of South Florida, Tampa, FL 33620
[2]Shanghai Institute of Ceramics, Chinese Academy of Sciences, Shanghai 200050, China

ABSTRACT

Dimensional nanocomposites of PbTe with varying carrier concentrations were prepared from undoped and Ag doped PbTe nanocrystals synthesized utilizing an alkaline aqueous solution-phase reaction. The nanocrystals were densified by Spark Plasma Sintering (SPS) for room temperature resistivity, Hall, Seebeck coefficient measurements, and temperature dependent thermal conductivity measurements. The nanocomposites show an enhancement in the thermoelectric properties compared to bulk PbTe with similar carrier concentrations, thus demonstrating a promising approach for enhanced thermoelectric performance.

INTRODUCTION

Thermoelectric phenomena allow the solid-state interconversion of thermal to electrical energy, offering another route to provide a sustainable supply of energy for the world's growing population. The effectiveness of a thermoelectric material is characterized by the thermoelectric figure of merit, $ZT=S^2T/\rho\kappa$, where S is the Seebeck coefficient, ρ is the electrical resistivity, T is the absolute temperature, and $\kappa = \kappa_L + \kappa_e$ is the total thermal conductivity with κ_L as the lattice contribution and κ_e as the electronic contribution [1]. In recent years, both theoretical and experimental studies have strongly suggested that large improvements in ZT could be achieved in nanostructured systems [2]. At the same time thermoelectric devices necessitate materials in large quantities, and practical synthesis techniques are essential to incorporate nanoscale features within a bulk material. Currently, the most common top down approach is through ball-milling bulk powders into nano-scale grains [3-6]. However, as suggested in the work of Heremans et al. [4], this technique can cause strain within these materials that directly affects their transport properties in an unpredictable manner. A study by Heremans et al. [7] demonstrates how PbTe with small concentrations of Pb nanograins can be produced. This led to an adoption of solution-phase synthesis technique for PbTe nanocomposites, which may minimize the above mentioned issue and lead to higher thermoelectric performance than that of the bulk.

This article reports a solution phase synthesis approach for preparing undoped as well as Ag doped PbTe nanocrystals that are then used to create bulk polycrystalline dimensional nanocomposites by spark plasma sintering (SPS). The Seebeck coefficient for the nanocomposites is larger than that of the bulk with similar carrier concentrations without a large increase in resistivity, thereby resulting in an enhanced thermoelectric figure of merit.

EXPERIMENTAL DETAILS

A low temperature reaction of a tellurium alkaline aqueous solution with a lead acetate trihydrate solution suitably precipitated out PbTe nanocrystals [8-9]. Various syntheses were conducted in order to optimize experimental conditions to produce ~100 nm spherical PbTe nanocomposites. Silver doped PbTe was prepared by adding silver acetate to the solution of lead acetate trihydrate before combining with the tellurium alkaline aqueous solution. Transmission electron microscopy confirmed the shape and size of the nanocrystals. This synthesis method yielded over 2g per batch, and can be scaled up for production of this thermoelectric material. Powder X-ray diffraction (XRD) validated the phase purity of the specimens. In traditional hot press techniques conglomeration of the nanocrystals is observed [10]. This non-homogeneity throughout the specimen is likely to alter the transport properties of the specimen in an unpredictable and unfavorable manner. To avoid conglomeration SPS was employed to densify the nanocrystals into bulk nanocomposites. The SPS procedure, which was carried out using a Sumitomo Dr. Sinter SPS-2040, consists of a pulsed DC current that conducts through the graphite die and the specimen under high pressure. This causes the specimen to be heated internally, providing a uniform and rapid thermal ramping while minimizing sintering time and temperature. Five polycrystalline PbTe nanocomposite specimens, named PbTe1-PbTe5 were prepared. Specimens PbTe1 and PbTe2 were nominally undoped and specimens PbTe3-PbTe5 were doped with Ag. The SPS technique successfully consolidated the nanoscale grains within a dense PbTe matrix with ~95% of theoretical density for each specimen.

RESULTS AND DISCUSSIONS

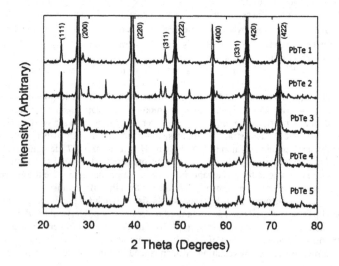

Figure 1: XRD spectrum of the undoped and Ag doped PbTe nanocomposites, indicating nearly phase pure specimens. The unassigned diffraction peaks are trace amounts of oxide impurities.

Figure 1 shows XRD spectra for the PbTe nanocomposites. Overall, the sharp peaks indicate the high crystallinty of the prepared specimens. The preservation of the nanostructures after the SPS densification was confirmed from scanning electron microscope (SEM) images of fractured surfaces. The size of the grains of the nanocomposites range from 100 nm to over 1 μm and minimal conglomeration of the nanoparticles was also observed.

The densified nanocomposite specimens were cut into 2 x 2 x 5 mm^3 parallelepipeds for transport property measurements. All the specimens were found to be p-type from room temperature Hall measurements, which was done by a four-probe method with a ±10% uncertainty. The measurement was conducted at both positive and negative magnetic fields in order to eliminate probe misalignment and temperature drift effects. Four-probe bipolar resistivity and gradient sweep Seebeck coefficient were measured using a custom vacuum probe with uncertainties of 4% and 6%, respectively. Table I shows the values of the room temperature physical properties for the five specimens. The carrier concentrations increased in the nanocomposite specimens with Ag doping, as listed in Table I.

Table 1: Room temperature ρ, S, κ and carrier concentration (p) values for the nanocomposite specimens.

Specimen	ρ (mOhm cm)	S (μV/K)	S^2/ρ (μW cm^{-1} K^{-2})	κ (Wm^{-1}K^{-1})	p (cm^{-3})
PbTe1	24.9	328	4.32	2.2	9.5 X 10^{17}
PbTe2	12.6	324	8.33	2.5	1.5 X 10^{18}
PbTe3	3.9	198	10.1	2.8	5.1 X 10^{18}
PbTe4	5.0	211	8.90	2.7	6.1 X 10^{18}
PbTe5	2.9	207	14.8	2.7	6.2 X 10^{18}

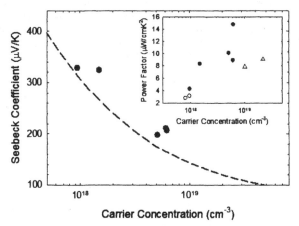

Carrier Concentration (cm^{-3})

Figure 2: Room temperature Seebeck coefficient versus carrier concentration for five polycrystalline p-type PbTe nanocomposites and the theoretically calculated relationship for the bulk (the dashed line) [11]. Inset: room temperature power factor as a function of carrier concentration for the five nanocomposites in comparison to bulk undoped PbTe (open circles) and bulk Na-doped PbTe (open triangles).

73

Figure 2 shows S of the five polycrystalline p-type PbTe nanocomposites together with a theoretical calculation for that of the bulk [11]. The S values are also given in Table I. The S values for the nanocomposites are larger than that of the bulk, with an enhancement of up to 23%, while the ρ values are not significantly higher as compared to that of the bulk. The nanocomposite with the best thermoelectric properties resulted in an enhanced room temperature S^2/ρ of 30% over that of the bulk [12]. The increased S^2/ρ (Table I) along with a similar thermal conductivity in the nanocomposites results in enhanced room temperature ZT of a factor of ~2 as compared to bulk PbTe. Temperature dependent ρ and mobility measurements on p-type PbTe nanocomposites, [12,13] together with theoretical calculations describing diffusion transport of carriers through a material with nanogranular regions [14], indicate unique temperature dependences as compared to that of the bulk. The results suggest an additional scattering mechanism in combination with phonon-carrier scattering dominant in single crystal and bulk polycrystalline lead chalcogenides. In the nanocrystalline materials, the trapping of carriers at grain boundaries may form energy barriers that impede the conduction of carriers between grains. This conduction can be effectively described as dominated by grain-boundary potential barrier scattering, in combination with phonon scattering. Furthermore, these nanocomposites demonstrate an enhanced thermoelectric performance as compared to bulk PbTe, suggesting interfacial energy barrier carrier scattering is an effective method of thermoelectric performance enhancement in bulk nanocomposites.

The thermal conductivity was measured from 300 K–12 K using the steady-state gradient sweep method in a custom radiation-shielded vacuum probe, with a maximum uncertainty of 8 % at 300 K. Figure 3 shows the temperature dependence of the total thermal conductivity for the nanostructured PbTe specimens. The estimated carrier contribution (κ_e) to the total thermal conductivity, calculated using the Wiedemann-Franz relation ($\kappa_e = L_0 T/\rho$, where the Lorenz number $L_0 = 2.45$ x 10^{-8} V^2/K^2), indicates a lattice-dominated thermal conduction with a negligible κ_e contribution to the total thermal conductivity, κ, due to the large resistivity (Table I). The temperature dependence of κ for all specimens is similar to bulk polycrystalline PbTe previously reported [15-19] with a maximum occurring between 20 K and 30 K and with similar room temperature values. These features are indicative of crystalline temperature dependence rather than the glass-like thermal conductivity observed in some nanostructured systems [3-7]. Therefore, the addition of nanostructure (100 nm grains) in these compounds does not strongly influence the thermal conductivity. The enhancement in the ZT is therefore entirely due to an increase in the S^2/ρ.

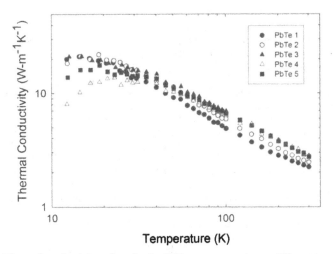

Temperature (K)

Figure 3: Thermal conductivity values for the PbTe nanocomposites at different temperatures.

CONCLUSIONS

PbTe nanocomposites were prepared by densifying ~100 nm PbTe nanocrystals synthesized in high yield employing a solution-phase technique. SPS successfully consolidated these nanoscale grains within a dense PbTe matrix. The carrier concentrations were modified by directly doping the PbTe nanocrystals with Ag prior to densification. The SPS densification technique in essence disperses the 100 nm nanocrystals with a bulk matrix. The results show an enhancement in the thermoelectric properties for the nanocomposites as compared to that of the bulk. This is the most direct evidence for ZT enhancement, as high as a factor of 2, due to the increase in the S^2/ρ.

ACKNOWLEDGEMENTS

The authors acknowledge support by the US Army Medical Research and Material Command under Award No. W81XWH-07-1-0708. Opinions, interpretations, conclusions, and recommendations are those of the authors and are not necessarily endorsed by the U.S. Army.

REFERENCES

[1] G. S. Nolas, J. W. Sharp, and H. J Goldsmid. *Thermoelectrics: Basic Principles and New Materials Developments,* (Springer-Verlag, Heidelberg, 2001), p. 45.
[2] S. V. Faleev and F. Léonard, Phys. Rev. B **77**, 214304 (2008).
[3] K. Kishimoto and T. Koyanagi. J. Appl. Phys. **92**, 2544 (2002).
[4] J. P. Heremans, C. M. Thrush, and D. T. Morelli. Phys. Rev. B. **70**, 115334 (2004).
[5] M. S. Dresselhaus, G. Chen, M. Y. Tang, R. G. Yang, H. Lee. D. Z. Wang, Z. F. Ren, J. P. Fleurial, and P. Gogna. Mater. Res. Soc. Symp. Proc. **886**, 3 (2006).
[6] B. Poudel, Q. Hao, Y. Ma, Y. Lan, A. Minnich, B. Yu, X. Yan, D. Wang, A. Muto, D. Vashaee, X. Chen, J. Liu, M. S. Dresselhaus, G. Chen, Z. Ren1, Science **320**, 634 (2008).
[7] J. P. Heremans, C. M. Thrush, and D. T. Morelli. J. Appl. Phys. **98**, 063703 (2005).
[8] W. Zhang, L. Zhang, Y. Cheng, Z. Hui, X. Zhang, Y. Xie, and Y. Qian, Mater. Res. Bull. **35**, 2009 (2000).
[9] C. B. Murray, C. R. Kagan, and M. G. Bawendi, Annu. Rev. Mater. Sci. **30**, 545 (2000).
[10] H. X. Xin and X. Y. Qin, J. Phys. D, Appl. Phys. **39**, 5331 (2006).
[11] A. J. Crocker and L. M. Rogers, Br. J. Appl. Phys. **18**, 562 (1967).
[12] J. Martin and G.S Nolas. Appl. Phys. Lett. **90**, 222112 (2007).
[13] J. Martin, L. Wang, L. Chen, and G. S. Nolas, Phys. Rev B **79**, 115311 (2009).
[14] L.M. Woods, A. Popescu, J. Martin, and G.S. Nolas, Mater. Res. Soc. Symp., current volume.
[15] W. Scanlon, Solid State Phys. **9**, 122 (1959).
[16] Z. H. Dughaish, Physica B **322**, 205 (2002).
[17] E. H. Putley, Proc. Phys. Soc. London, Sect. B **65**, 388 (1952).
[18] E. H. Putley, Proc. Phys. Soc. London, Sect B **65**, 736 (1952).
[19] Yu. I. Ravich, B. A. Efimova, and I. A. Smirnov, *Semiconducting Lead Chalcogenides* (Plenum, New York, 1970), Vol. 5, p.91.

Mater. Res. Soc. Symp. Proc. Vol. 1166 © 2009 Materials Research Society 1166-N04-01

Understanding Electrical Transport and the Large Power Factor Enhancements in Co-Nanostructured PbTe

Joseph R. Sootsman[1], Vladimir Jovovic[2], Christopher M. Jaworski[2], Joseph P. Heremans[2], Jiaqing He[3], Vinayak P. Dravid[3], Mercouri G. Kanatzidis[1]

1. Department of Chemistry, Northwestern University, Evanston, IL 60208
2. Department of Mechanical Engineering, Ohio State University, Columbus, OH 43210
3. Department of Materials Science, Northwestern University, Evanston, IL 60208

ABSTRACT

We previously reported the synthesis of nanostructured composite PbTe with excess Pb and Sb metal inclusions. The electrical conductivity shows an unusual temperature dependence that depends on the inclusion Pb/Sb ratio, resulting in marked enhancements in power factor and ZT at 700 K. Additional investigation of the transport and structure of these materials is reported here. Measurements of the scattering parameter reveals there is little change in electron scattering with respect to pure PbTe. High resolution electron microscopy was used to determine additional information about the nature of the precipitate phases present in the samples. High temperature transmission electron microscopy reveals that the precipitates begin to dissolve at high temperatures and completely disappear at T > 619K. A qualitative explanation of the unusual transport behavior of these materials is presented.

INTRODUCTION

Emerging trends from thermoelectric research indicate that nanostructured materials can achieve a high figure of merit.[1-3] The figure of merit, which determines thermoelectric device conversion efficiency, is related to the transport properties of a material through $ZT=S^2\sigma T/\kappa$ where S is the Seebeck coefficient, σ is the electrical conductivity, κ the thermal conductivity, and T the temperature.[4] Nanostructured materials can have increased ZT because of low thermal conductivity resulting from phonon scattering at the interfaces of nanoscale features.[5]

Recently, we reported the synthesis and thermoelectric properties of PbTe nanostructured with nanoscale precipitates of Pb and Sb.[6] These nanoprecipitates profoundly impact the temperature dependence of the mobility. Typically in degenerately doped PbTe the mobility decreases by a power law relationship ($\mu=aT^\alpha$) where the exponent (α) is -2.2 to -2.5. This is a result of acoustic phonon scattering of the electrons at high temperature. In samples with both Pb and Sb nanoinclusions this power law relationship is no longer valid and in several cases the mobility actually increases before it falls at high temperature. This behavior is tuned by the Pb/Sb ratio.[6] In these materials the lattice thermal conductivity is also tuned by the Pb/Sb ratio where samples rich in Sb tend to have a reduced thermal conductivity. Here we present additional transport measurements and transmission electron microscopy (TEM) results that add to our understanding of the novel behavior of these materials. First, measurements were performed on several samples using the method of 4 coefficients in order to determine the scattering parameter.[7] The scattering parameter is related to the temperature dependence of the mobility and Seebeck coefficient. These measurements indicate that the high temperature electron scattering is similar to that of PbTe itself. TEM investigations of these materials at high temperature indicate the precipitates may begin to dissolve and disappear at 619K. From these results a possible mechanism is presented to explain the transport behavior observed.

EXPERIMENTAL

PbTe was prepared without additional dopants using high purity metals on the 200-300g scale in evacuated silica ampoules by reacting 149.4931 g of Pb and 92.0629 g of Te at 1000° C for 12 h. Further reaction of this starting material was performed by the addition of metallic Pb and Sb to PbTe in the appropriate molar ratios (actuing as n-type dopants). The sealed tubes were then fired at 1000°C allowing the mixture to homogenize over 10-15hrs. The tubes were then rapidly cooled in a room temperature water bath. The resulting ingots were strong and were easily cut for transport and microscopic characterization.

Scattering parameter measurements were performed according to reference 7. High resolution transmission electron microscopy was performed on samples prepared by conventional methods using a JEOL 2100F electron microscope. Heating experiments were performed from room temperature to 700K using a Gatan sample holder.

DISCUSSION

Temperature dependent transport measurements were performed on several samples to determine the electrical conductivity, Seebeck coefficient, and Hall and Nernst coefficients, Figure 1a,b,d,e.[7] The magnitude of the electrical conductivity in these samples was somewhat higher than those measured previously[6], possibly a result of variation of the rapidly cooled materials as well as in the measurement itself. Large power factors ($\sim30\mu W/cmK^2$) were again observed at T>600K, Figure 1c. From these measurements the scattering parameter, carrier concentration, Lorenz number and effective mass can be calculated.

The scattering parameter (λ) is related to the relaxation time in solids with parabolic bands by, $\tau = \tau_o E^{\lambda-1/2}$, where τ is the relaxation time and E the energy. In turn the electrical conductivity and Seebeck coefficient are also dependent on the relaxation time. In pure PbTe the scattering parameter is ~0.5 indicative of acoustic phonon scattering. In samples of PbTe with Pb nanoprecipitates the scattering parameter was found to be much higher than in pure PbTe, increasing the Seebeck coefficient by enhancing the energy dependence of the relaxation time.[7] In samples of PbTe with Pb and Sb inclusions the extracted λ from the transport data was found to be between 0 and 0.8 over the entire temperature range measured (Figure 1f), similar to pure PbTe. This result is rather surprising given the large changes in the behavior of the temperature dependence of the electrical conductivity. It is also interesting to note that the Seebeck and Hall coefficients vary smoothly with temperature and do not show any anomalous behavior, indicating that the carrier concentration appears to be stable. If neither the λ nor the carrier concentration is changing, another mechanism must be responsible for the anomalous behavior of the mobility and enhancement of the power factor.

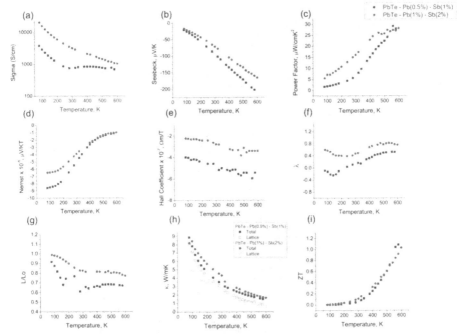

Figure 1. Temperature dependent measurement of (a) Electrical conductivity, (b) Seebeck coefficient, (c) Power factor, (d) Nernst coefficient, (e) Hall coefficient, (f) scattering parameter λ, (g) Lorenz ratio (h) thermal conductivity, and (i) ZT for two PbTe samples nanostructured with Pb and Sb.

The Lorenz number was also calculated from the transport data, Figure 1g. At room temperature a ratio of ~1 between L/L_o indicates that the electronic thermal conductivity can accurately be determined by the Widemann-Franz law, however at high temperatures there is significant deviation from L_o. Using the experimentally determined Lorenz number and the measured thermal conductivity at low temperatures (steady state method) and high temperatures (Flash Diffusivity – Heat Capacity method) the lattice thermal conductivity was calculated, Figure 1h. Again, very low lattice thermal conductivity (~0.5 W/mK) values were obtained and ZT values of ~1 were measured at 600K for these compositions, Figure 1i.

The ZT values measured represent a significant increase compared to similarly doped PbTe and can be attributed to both the large power factor and reduced thermal conductivity. For example, the carrier concentration of the PbTe-Pb(0.5%)-Sb(2%) sample is 9.95 x 10^{18} cm^{-3} at 600K and the power factor is 28μW/cmK^2. For similarly doped PbTe the power factor is only 19 μW/cmK^2 at 600K. This increase, combined with the ~50% decrease in lattice thermal conductivity, results in a 40% increase in ZT at this temperature. In order to better understand and perhaps model this composite, additional temperature dependent understanding of the precipitates and their structure is necessary.

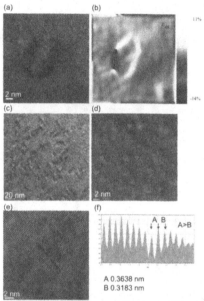

A 0.3638 nm
B 0.3183 nm

Figure 2. High Resolution transmission electron micrograph showing one Sb nanoprecipitate in the PbTe-Sb(1.5%) sample and (b) its corresponding strain calculated with reference to the PbTe matrix. (c-d) Low magnification images of PbTe-Pb(2%) showing two types of precipitates present including spherical and straight line defects. One straight line defect (e) and its corresponding line scan (f) indicates this defect may be deficient in Te.

At room temperature it is expected that the PbTe-Pb-Sb composite material contains both Pb and Sb precipitates as these elements have a eutectic phase relationship and the solubility of excess Pb and Sb is low in PbTe. In order to better understand the nature of the Pb and Sb precipitates respectively the single precipitate systems were investigated in more detail. The PbTe-Sb(1.5%) material shows precipitates with a large amount of strain at their interface with PbTe, Figure 2a-b. It is likely this strain is responsible for the large reduction of thermal conductivity observed. The PbTe-Pb system reveals there may actually be two types of precipitates present in the PbTe-Pb material. Both spherical precipitates and "line" type defects were observed. The "line" type defects appear to be deficient in Te. Little strain was observed at the boundary between the PbTe and Pb precipitates. The nature of these precipitates may be useful for future modeling of the complex composites. Additional investigation of the Pb and Sb precipitates as a function of temperature may also be useful for these models.

Transmission electron microscopy was performed using a heating stage in order to determine the stability of the Pb and Sb precipitates at high temperature. A constant heating rate of 10/min was used and images were collected from room temperature to 678K, Figure 3. A hole was created in the sample to accurately determine the location of the sample (bright spot in images). Initially, at room temperature a large number of precipitates are clearly observed. As the temperature was increased, the precipitates begin to dissolve at approximately 543-578K. At higher temperatures the precipitates continue to dissolve and eventually completely disappear at

678K.[8] As the sample was cooled the precipitates reappear and are present at room temperature. The unusual temperature dependence of the mobility mentioned above is likely related to the behavior of electrons at the interfaces as well as within the nanoprecipitates. The TEM results lead to the following hypothesis for the observed mobility behavior. At room temperature the PbTe material contains Pb and Sb precipitates. It was shown previously that Pb precipitates significantly reduce the electron mobility.[7] As the temperature increases the Pb precipitates that significantly reduce the mobility begin to dissolve and the mobility increases. As the temperature increases further acoustic phonon scattering again dominates and the mobility falls with increasing temperature. It is unclear why the dissolution of the precipitates into the PbTe matrix does not significantly alter the carrier concentration. Theoretical calculations utilizing the temperature dependent structural characterization may provide sufficient models to understand the behavior of this composite.

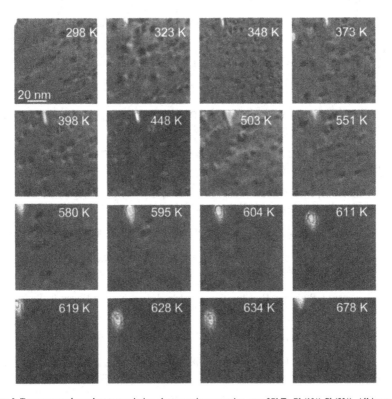

Figure 3. Temperature dependent transmission electron microscope images of PbTe-Pb(1%)-Sb(2%). All images have the same magnification. The bright spot at the top left of the image is a result of intentional electron damage to track the location of the sample during heating.

CONCLUSIONS

Transport measurements and transmission electron microscopy analysis of co-nanostructured PbTe with Pb and Sb precipitates have been presented. It has been determined that the scattering parameter is not significantly altered in these materials, although the exact mechanism of the transport is still under investigation. The nanostructures within the composite were identified as "line" type defects deficient in Te, as well as spherical nanoprecipitates of Sb and Pb. At high temperature these precipitates appear to dissolve in thin foils during TEM measurements. The structure and temperature dependent behavior of the nanoprecipitates may be useful for future studies.

ACKNOWLEDGMENTS

Financial support from the Office of Naval Research is gratefully acknowledged. Portions of the transmission electron microscopy work was performed in the (EPIC) (NIFTI) (Keck-II) facility of NUANCE Center at Northwestern University. NUANCE Center is supported by NSF-NSEC, NSF-MRSEC, Keck Foundation, the State of Illinois, and Northwestern University.

REFERENCES

1.	Hsu, K. F.; Loo, S.; Guo, F.; Chen, W.; Dyck, J. S.; Uher, C.; Hogan, T.; Polychroniadis, E. K.; Kanatzidis, M. G. *Science* **2004**, *303*, 818-821.

2.	Androulakis, J.; Hsu, K. F.; Pcionek, R.; Kong, H.; Uher, C.; D'Angelo, J. J.; Downey, A.; Hogan, T.; Kanatzidis, M. G. *Advanced Materials* **2006**, *18*, 1170-1173.

3.	Androulakis, J.; Lin, C.-H.; Kong, H.-J.; Uher, C.; Wu, C.-I.; Hogan, T.; Cook, B. A.; Caillat, T.; Paraskevopoulos, K. M.; Kanatzidis, M. G. *J. Am. Chem. Soc.* **2007**, *129*, 9780-9788.

4.	Rowe, D. M. *CRC handbook of thermoelectrics*; CRC Press: Boca Raton, FL, 1995.

5.	Kim, W.; Singer, S. L.; Majumdar, A.; Zide, J. M. O.; Klenov, D.; Gossard, A. C.; Stemmer, S. *Nano Lett.* **2008**, *8*, 2097-2099.

6.	Sootsman, J. R.; Kong, H.; Uher, C.; D'Angelo, J. J.; Wu, C.-I.; Hogan, T. P.; Caillat, T.; Kanatzidis, M. G. *Angew. Chem., Int. Ed. Engl.* **2008**, *47*, 1-6.

7.	Heremans, J. P.; Thrush, C. M.; Morelli, D. T. *J. Appl. Phys.* **2005**, *98*.

8.	It should be noted that the dynamics of the precipitates of Pb and Sb in the bulk may not be accurately described by the behavior in thin foils like those used in TEM investigation

Mater. Res. Soc. Symp. Proc. Vol. 1166 © 2009 Materials Research Society   1166-N04-03

Enhancement of Thermoelectric Figure-of-Merit by a Nanostructure Approach

Z. F. Ren[1], B. Poudel[2], Y. Ma[1], Q. Hao[3], Y. C. Lan[1], A. Minnich[3], A. Muto[3], J. Yang[1], B. Yu[1], X. Yan[1], D. Z. Wang[1], J. M. Liu[4], M. S. Dresselhaus[5], and G. Chen[3]

[1]Department of Physics, Boston College, Chestnut Hill, Massachusetts 02467, U.S.A.
[2]GMZ Energy, Inc., Waltham, Massachusetts 02453, U.S.A.
[3]Department of Mechanical Engineering, MIT, Cambridge, Massachusetts 02139, U.S.A.
[4]Department of Physics, Nanjing University, Nanjing, China
[5]Department of Physics, MIT, Cambridge, Massachusetts, 02139, U.S.A.

ABSTRACT

The dimensionless thermoelectric figure-of-merit (ZT) in bulk materials has remained about 1 for many years. Here we show that a significant ZT improvement can be achieved in nanocrystalline bulk materials. These nanocrystalline bulk materials were made by hot-pressing nanopowders that are ball-milled from either crystalline ingots or elements. Electrical transport measurements, coupled with microstructure studies and modeling, show that the ZT improvement is the result of low thermal conductivity caused by the increased phonon scattering by grain boundaries and defects. More importantly, the nanostructure approach has been demonstrated in a few thermoelectric material systems, proving its generosity. The approach can be easily scaled up to multiple tons. Thermal stability studies have shown that the nanostructures are stable at the application temperature for an extended period of time. It is expected that such enhanced materials will make the existing cooling and power generation systems more efficient.

INTRODUCTION

For several decades, thermoelectric devices have attracted extensive interest because of their excellent features: no moving parts, quiet operation, environmentally friendly, high reliability, etc [1-5]. These solid-state thermoelectric devices based on thermoelectric effects can be used in a wide range of applications in waste heat recovery, air-conditioning, and refrigeration. The efficiency of thermoelectric devices is determined by the materials' dimensionless figure-of-merit (ZT), defined as $ZT = (S^2\sigma/k)T$, where S, σ, k, and T are the Seebeck coefficient, electrical conductivity, the total thermal conductivity made of contributions from the lattice (k_L) and the electrons (k_e), and absolute temperature, respectively [1-3]. To make a thermoelectric device competitive, an average ZT in the application temperature range higher than 1 is required [1-3].

There have been persistent efforts to improve ZT since the 1950's, but the peak ZT of dominant commercial materials based on Bi_2Te_3 and its alloys, such as $Bi_xSb_{2-x}Te_3$ (p-type), has remained at 1. One traditional way to improve ZT is to reduce the lattice thermal conductivity k_L via alloying, which does not significantly reduce the electrical properties. Over the last decade, several groups have reported enhanced ZT in superlattices such as Bi_2Te_3/Sb_2Te_3 [6] and $PbSe_{0.98}Te_{0.02}/PbTe$ [7], and in new bulk materials, such as silver antimony lead telluride (LAST) and its alloys [8]. Although high ZT values were reported in superlattice structures, it is difficult to use them in large scale energy conversion applications. This is mainly due to limitations in both heat transfer and cost. Bulk materials with improved ZT such as LAST are ideal for high temperature operations. However, at near room temperature (0-250 °C), Bi_2Te_3-based materials are still the dominant players.

We are pursuing a new approach in which the primary cause of ZT enhancement in superlattices, reduced thermal conductivity, also exists in random nanostructures [9-13]. Here, we report a significant ZT increase in nanocrystalline bulk p-type $Bi_xSb_{2-x}Te_3$, reaching a peak ZT of 1.4 at 100 °C. The enhanced ZT is the result of a significant reduction in thermal conductivity due to strong phonon scattering by nano grains in the nanostructures. There have also been reports of ZT improvements at room temperature in Bi_2Te_3-based materials caused by the addition of Bi_2Te_3-nanotubes [14] and melt spinning [15]. Our method, on the other hand, is based on the ball milling and hot-pressing of nanoparticles. This approach is simple, cost effective, and can be used on other materials. Our materials have a ZT of about 1.2 at room temperature and 0.8 at 250 °C with a peak 1.4 at 100 °C. In comparison, conventional Bi_2Te_3-based materials have a peak ZT of about 1 at room temperature and about 0.25 at 250 °C. The high ZT in the temperature ranges of 25 – 250 °C makes the nanocrystalline bulk materials attractive for cooling and low-grade waste heat recovery applications. The materials can also be integrated into segmented thermoelectric devices for thermoelectric power generation which operate at high temperatures. In addition to the high ZT values, the nanocrystalline bulk materials are also isotropic. They do not suffer from the cleavage problem that is common in traditional zone-melting made ingots, thus leading to easier device fabrication and system integration, and a potentially longer device lifetime.

EXPERIMENTAL

Nanopowders were made by ball milling bulk p-type BiSbTe alloy ingots or by ball milling elemental chunks of Bi, Sb, and Te into alloy nanopowders. P-type BiSbTe alloy ingots or pure elements were loaded into a jar with balls as milling media inside the argon filled glove box to avoid any oxidation. The jar was loaded into a ball mill and processed for a number of hours. When the as-prepared nanopowders were ready, the nanopowders were then loaded into a graphite die with an inner diameter of 12.7 mm in the glove box, and hot-pressed into 100% dense solid nanocrystalline bulk disk samples using a direct current (DC) induced hot press in air. Disks of ½″ (diameter) and 2 mm thickness, and bars of about $2 \times 2 \times 12$ mm^3, were cut from the hot pressed bulk disks along the axial or disk plane directions. The cut disks and bars were polished for electrical and thermal conductivity and for the Seebeck coefficient measurements.

The as-prepared nanopowders and hot pressed bulk samples were characterized by X-ray diffraction (Bruker-AXS) using Cu Kα radiation, field emission scanning electron microscopy (JEOL-6340F), and transmission electron microscopy (JEOL-2010F).

The electrical conductivity was measured by a four-point direct current current-switching technique. Meanwhile, the Seebeck coefficient S was measured by a static DC method based on the slope of a voltage vs. temperature-difference curve, using a commercial equipment (ZEM-3, Ulvac, Inc.) on the same bar sample with a dimension of 2×2 mm^2 in cross-section and 12 mm in length. The properties in the same sample were also measured by a homemade system and the two sets of measurements are within 5% of each other. The thermal diffusivity (α) was first measured by a laser-flash method on a disk using a commercial system (LFA 447 Nanoflash, Netzsch Instruments, Inc.). After the measurement, bars were diced from the disks and their thermal diffusivities were measured using the Ångström method in the same home-built system. The thermal diffusivity α values of the bar and of the disk are in an agreement range of 5%. Specific heat c_p of the materials was determined by a commercial differential scanning calorimeter (DSC 200-F3, Netzsch Instruments, Inc.). The thermal conductivity k was calculated using the equation $k = \alpha \times \rho \times c_p$, where ρ is the density of samples.

The structures of the hot-pressed nanocrystalline samples were investigated on a JEOL-2010F transmission electron microscope (TEM) operated at 200kV. The hot-pressed nanocrystalline bulk samples were cut into blocks of $2 \times 3 \times 1$ mm^3 and mechanically polished down to $2 \times 3 \times 0.02$ mm^3 using a mechanical Tripod Polisher (South Bay Technology Inc.). The polished samples were glued to copper grids and Ar$^+$-ion milled to electron transparent using a precision ion polishing system (Gatan Inc.) for less than 30 minutes with incident energy of 3.2 kV and beam current of 15 μA at an incident angle of 3.5 degrees.

RESULTS

Ball Milling of Commercial Ingot, Hot Press, and Characterizations

In the beginning, bulk p-type BiSbTe alloy ingots were ball milled into nanopowders and the as-prepared nanopowders were compacted into nanocrystalline bulk samples by hot-press.

Figure 1 shows the x-ray diffraction (XRD) pattern of the nanopowders after ball milling (Figure 1A), scanning electron microscope (SEM) image (Figure 1B), and low- and high-magnification transmission electron microscope (TEM) images (Figures. 1C and 1D). The XRD

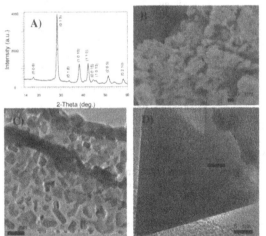

pattern verifies that the powder is in a single phase and is well matched with $Bi_{0.5}Sb_{1.5}Te_3$. The broadened diffraction peaks indicate that the particles are small, which is also confirmed in the SEM image (Figure 1B) and the low-magnification TEM image (Figure 1C). The TEM image (Figure 1C) also shows that the nanoparticles have sizes ranging up to 50 nm, with an average size of about 20 nm. The high resolution TEM (HRTEM) image (Figure 1D) confirms the excellent crystallinity of the nanoparticles and the clean surfaces. The inset in Figure 1D also shows that some of the nanoparticles are even smaller than 5 nm.

Figure 1. XRD (A), SEM (B), low- (C) and high- (D) Magnification TEM images of an as-ball-milled nanopowder.

85

Microstructure of the hot-pressed nanocrystalline bulk samples were studies by TEM. Figure 2A is a typical bright-field TEM (BF-TEM) image of four grains in the nanocrystalline bulk. There is no cavity between grains, consistent with 100% density in the nanocrystalline bulk. Usually the grains are polygonal. Selected area electron diffraction shows that the grains are single crystals with random orientation. Figure 2B shows an HRTEM image of grain boundaries between three adjacent grains. The high-angle grain boundaries are well crystallized and free of second phase. The crystalline orientation randomness of the adjacent grains helps phonon scattering significantly. Figure 2C indicates an HRTEM image of smaller grains. These small grains are also single crystals and closely packed, similar to the bigger grains shown in Figure 2A. Figure 2D shows the grain size distribution. Clearly, the majority of the grains are of diameters below 1.0 μm. The inset of Figure 2D shows that 12% of the grains having a diameter 0 – 20 nm and 5% having a diameter 20 – 40 nm. The grain boundaries scatter phonons with mean free path comparable to the grain size, reducing the thermal conductivity. A wide grain size distribution as shown in Figure 2D is responsible to scattering phonons with a variety of wavelength.

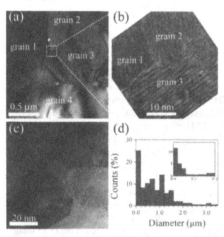

Figure 2. (A) BF-TEM image of multi grains, (B) HRTEM of grain boundaries, (C) HRTEM images of smaller grains, and (D) grain size distribution in the nanocrystalline bulk.

In addition to the clean boundaries between grains, as shown in Figures 2B and 2C, there are about 50% grains in the nanocrystalline bulk surrounded by nanometer thick interface regions. The energy dispersive X-ray spectrum (EDS) indicates that the interface region is slightly bismuth-rich, 1.0 ± 0.5 atomic% higher than that of the grains. This bismuth-rich region builds up charges, and thus increases the hole carrier concentration in the grains. At the same time, the interface region scatters phonons. Therefore, the bismuth-rich interface region blocks the phonon transportation while benefits the charge carrier transportation.

There are many three-dimensional nano-precipitates either embedded in the grains or located at the grain boundaries. In the experiments, four kinds of nano-precipitates in the nanocrystalline bulk are observed, and are shown in Figures 3A-3D respectively. One kind is nano-precipitate without boundaries (Figure 3A). There is no any obvious lattice distortion between the grain matrix and the nano-precipitate. The nano-precipitates are brighter than the surrounding crystalline grain matrix. EDS (curve *a* in Figure 3E) shows the nano-precipitate is antimony-rich and tellurium-poor. The second type, as shown in Figure 3B, is nanodot with twisted boundaries. The chemical composition of the nanoprecipitates is the same as the surrounding matrix. The third type, as shown in Figure 3C, has also twisted grain boundary with the surrounding matrix but with antimony-rich. EDS of such a nano-precipitate with Sb-rich is plotted as curve *c* in Figure 3E. The fourth type is pure tellurium nanodot, shown in Figure 3D. EDS of such a precipitate is plotted as curve *d* in Figure 3E. The pure tellurium

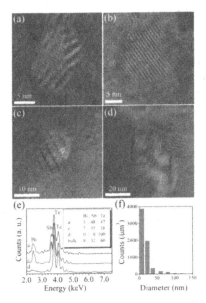

(e) EDS spectra panel axes: Counts (a. u.) vs Energy (keV) 2.0 3.0 4.0 5.0 6.0 7.0

(f) histogram panel axes: Counts (μm) vs Diameter (nm)

Figure 3. HRTEM images of nano-precipitates embedded in the grain matrix. (A) an antimony-rich precipitate without boundary, (B) a precipitate with twisted boundary but the same composition of the surrounding matrix, (C) an antimony-rich precipitate with twisted boundary, (D) a tellurium precipitate with high-angle boundary. (E) EDS spectra of the precipitates in (A)-(D). (F) precipitate size distribution.

precipitates are polygonal while other three kinds of precipitates are irregular. Because the chemical composition of the antimony-rich precipitates and tellurium precipitates are different from that of the surrounding grain matrix, the precipitates would increase the carrier concentration in the grains, just as bismuth-rich interface regions do. Hall effect measurements at room temperature confirm the fact that the hole carrier concentration of the nanocrystalline bulk (n = 2.5×10^{19}/cm^3) is about 39% higher than that of the ingot (n = 1.8×10^{19}/cm^3).

Generally speaking, the nanoprecipitates are less than several ten nanometers. These nanoprecipitates probably form during hot-press heating and cooling processes. Similar types of nanoprecipitates have been observed in LAST alloys and were allegedly responsible for ZT enhancement in those alloys [8]. However, since there are so many interfaces from nanograins in our material, nanoprecipitates may not be the only reason for the strong phonon scattering. The larger-sized grains containing nanoprecipitates are likely due to the non-uniform milling of the ingot during ball milling, and may have experienced some grain growth during the hot-press compaction via Oswald Ripening. More uniform nanograins during ball milling may keep these nanograins from growing during the hot-press.

Figure 3F is the frequency curve of these precipitates. The precipitate concentration in the nanostructured bulk increases rapidly with decreasing precipitate size. The volume of these nano-precipitates accounts for ~5% of the nanocrystalline bulk. Besides increasing the carrier concentration, the nano-precipitates in nanograins scatter phonons, especially phonons with mean free path matched with the precipitate size, thus reducing lattice thermal conductivity.

Besides precipitates, there are other kinds of defects in the nanograins of the nanocrystalline bulks. TEM investigation indicates that the five-layer and seven-layer lamellae stack randomly along c-axis, forming stacking faults in the nanograins. Threading dislocation concentration is ~10^{11}/cm^2 in nanograins, at least ten times higher than ~5×10^9/cm^2 found in the crystalline ingot. The point defect concentration in nanograins is 2 – 3 orders of magnitude higher than that of the ingot. The structural modulations are also observed in the nanograins while the concentration is almost same as that in the ingot. All these defects with higher concentration would scatter phonons more effectively and further decrease thermal conductivity of the nanocrystalline bulks.

Figures 4 shows the temperature dependence of the electrical conductivity (Figure 4A), Seebeck coefficient (Figure 4B), power factor (Figure 4C), thermal conductivity (Figure 4D), and ZT (Figure 4E) of a typical nanocrystalline bulk sample in comparison with the state-of-the-art (SOA) p-type BiSbTe alloy ingot. All the properties are measured on the same direction and reproduced on many samples. Our findings show that while the electrical conductivity of the nanocrystalline bulk sample is slightly higher than the SOA ingot (Figure 4A), the Seebeck coefficient is either slightly lower or higher than the ingot depending on its temperature (Figure 4B). Ultimately, the power factor ($S^2\sigma$) values are comparable to that of an ingot below 50 °C and higher above 75 °C (Figure 4C). We also found that the thermal conductivity of the nanocrystalline bulk samples is significantly lower than that of the ingot and, more importantly, the difference increases with the increasing temperature (Figure 4D), leading to significantly enhanced ZTs (Figure 4E) in the temperature range of 20 – 250 °C. It also shows that the peak ZT value shifts to a higher temperature (100 °C). The peak ZT of nanocrystalline bulk samples is about 1.4 at 100 °C, which is significantly higher than the SOA Bi_2Te_3-based alloys. The ZT value of the SOA ingot starts to drop above 75 °C and is below 0.25 at 250 °C, whereas the nanocrystalline bulk samples are still higher than 0.8 at 250 °C. Such ZT characteristics are best for power generation applications due to a lack of available materials with high ZT in this temperature range. All of these measurements were confirmed by the two independent techniques on more than 100 samples.

In comparing the transport properties of the nanocrystalline bulk samples with the SOA ingot, it is important to note the slow increase in the thermal conductivity as a function of temperature (Figure 4D). This indicates a smaller bipolar contribution [2] to the conductivity by thermally generated electrons and holes in the nanocrystalline bulk materials. We explain this reduced bipolar effect by assuming the existence of an interfacial potential that scatters more electrons than holes. Past studies in Bi_2Te_3-related materials suggested that structural defects such as antisites, i.e., Bi atoms go to Te sites, serve as an important doping mechanism [16,17]. We anticipate that such antisites are

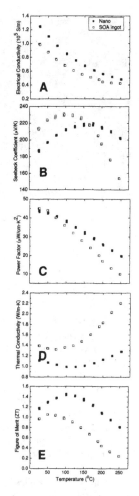

Figure 4. Temperature dependence of the electrical conductivity (A), Seebeck coefficient (B), power factor (C), thermal conductivity (D), and ZT (E) of a hot-pressed nanocrystalline bulk sample in comparison with an SOA ingot.

88

more likely to occur at interfaces. Uncompensated recombination centers at interfaces associated with defect states and antisites are responsible for charge buildup at grain-boundaries and thus increase the hole density in the grains. This is consistent with the observed increase in the electrical conductivity as well as the reduction in the Seebeck coefficient of the nanocrystalline bulk samples compared to those of the SOA ingot parent material (Figs. 4A and 4B).

Ball Milling of Elemental Chunks, Hot Press, and Characterizations

To simplify the process, it is better to eliminate the ingot processing step. Therefore, we studied the process using elemental chunks. To make nanopowders from elemental chunks, appropriate amounts of elements (Bi, Sb, and Te) with a nominal composition of $Bi_{0.4}Sb_{1.6}Te_3$ were loaded into a ball mill jar with balls, and then subjected to mechanical alloying.

Figure 5 shows the XRD pattern (Figure 5A), SEM image (Figure 5B), BF-TEM image (Figure 5C), and HRTEM image (Figure 5D) of the nanopowders after ball milling. The XRD pattern verifies that the powders are single phase, indicating that the mechanically assisted reaction during ball milling can make elemental chunks Bi, Sb and Te into single phase alloy. The broadened diffraction peaks indicate that the particles are small, which is also confirmed by the SEM image (Figure 5B) and BF-TEM image (Figure 5C). The BF-TEM image (Figure 5C) shows that the nanoparticles have sizes of about 5 to 20 nm with an average size of about 10 nm. The HRTEM image (Figure 5D) confirms the good crystallinity of the nanoparticles and clean surfaces.

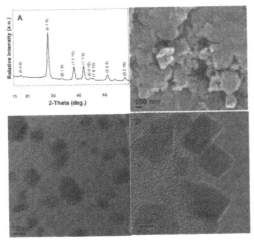

Figure 5. (A) XRD pattern, (B) SEM, (C) BF-TEM, and (D) HRTEM images of mechanically alloyed nanopowders after ball milling from elements.

The nanostructure of the hot-pressed nanocrystalline bulks from elemental chunk was also carried out on TEM. BF-TEM image (Figure 6A) indicates that one of the common features is the presence of submicrometer grains and dense packing. An HRTEM image (Figure 6B) shows that grains have excellent crystallinity and large angles between them, which explains isotropy of properties. The grains are grown to larger sizes in comparison with the starting nanopowders (Figure 5C). More TEM studies indicated that nanoprecipitates are also formed and embedded inside the BiSbTe nanograins, very similar to the hot-pressed nanocrystalline bulks made from ingot. These nanoprecipitates should be formed due to the composition fluctuation inside the grains. We believe that the size of the individual nanograins and nanoprecipitates all help to enhance the scattering of the phonons in different energy ranges and therefore contribute to the reduction of the thermal conductivity.

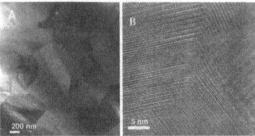

Figure. 6. (A) BF-TEM and (B) HRTEM of nanocrystalline bulks made from elements.

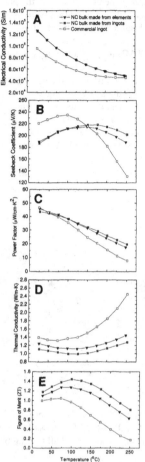

Figure 7. (A) Electrical conductivity, (B) Seebeck coefficient, (C) power factor, (D) thermal conductivity, and (E) ZT dependence of temperature of hot-pressed nanocrystalline bulk samples made from elemental chunks in comparison with a commercial ingot and a nanocrystalline bulk sample made from ingot.

In Figure 7, we compared the transport properties of three samples: the hot-pressed nanocrystalline sample made from elements, the hot-pressed nanocrystalline sample made from crystalline ingot, and the crystalline ingot. The behaviors of the two nanocrystalline samples are similar. The electrical conductivity of the nanocrystalline sample from elements is always higher than that of the crystalline ingot sample (Figure 7A) because of the higher carrier concentration (2.9×10^{19} cm^{-3} for nanocrystalline bulks made from elements and 1.8×10^{19} cm^{-3} for ingot), and is almost identical to the nanocrystalline samples made from ingot. The Seebeck coefficient (Figure 7B) of nanocrystalline samples from elements is slightly lower than that of nanocrystalline samples from ingot, both of which have a lower Seebeck coefficient than that of the crystalline ingot sample below 150 °C but higher above 150 °C. The smaller Seebeck coefficient near room temperature is due to higher carrier concentration while the larger Seebeck coefficient at higher temperatures is due to suppression of minority carrier (electron) excitation in more heavily doped samples. Figure 7C shows the corresponding power factor of the three samples. The nanocrystalline samples from ingots have a power factor comparable to that of the crystalline ingot below 100 °C but higher above 100 °C than that of the ingot, while the nanocrystalline sample from elements have slightly lower power factor than that made from ingot. Thermal conductivity of the nanocrystalline sample from elements is significantly lower than that of the ingot as expected due to the

90

increased phonon interface scattering, but systematically higher than that of the nanocrystalline samples made from ingot, probably due to the lack of some minor elements that were used in the ingot and some minor structural difference. The smaller thermal conductivity of the nanocrystalline samples at higher temperatures is due to weakened bi-polar effect resulted from the interface thermionic emission.

Figure 7E shows the temperature dependence of ZT for the hot-pressed nanocrystalline sample made from elements in comparison with the commercial ingot and the nanocrystalline bulk sample made from ingot. The peak ZT value shifts to a higher temperature and remains significantly higher than that of the ingot at all temperatures, but about 10 % lower than that of the nanocrystalline bulks made from ingot. The peak ZT of our hot pressed samples is about 1.3 at 100 °C, which is significantly higher than that of the best Bi_2Te_3-based alloy ingots. The superior ZT at high temperatures is very important for power generation applications since there are no other materials with a similar high ZT at this temperature.

Phonon Scattering in Nanocrystalline Bulks

According to the scattering theory, all imperfections in the nanocrystalline bulks disrupt the phonon path and decrease lattice thermal conductivity. The lattice thermal conductivity is related to a total scattering rate of phonons in the nanocrystalline bulks, which can be given by $\dfrac{1}{\tau_{total}(\omega)} = \dfrac{1}{\tau_{g,b}} + \dfrac{1}{\tau_i} + \dfrac{1}{\tau_p} + \dfrac{1}{\tau_d} + \dfrac{1}{\tau_{p,d}} + \dfrac{1}{\tau_U} + \dfrac{1}{\tau_e} + \cdots$, where the right terms are the phonon scattering rates for grain boundary, interface regions, precipitates, dislocations, point defects, Umklapp phonons, phonon-electron scattering, and other scatterings. These scatterings enhance the total phonon scattering rate of the nanocrystalline bulk. Compared with the ingot, the enhancement of the total scattering rate can be expressed as $\Delta\left(\dfrac{1}{\tau_{total}}\right) \approx \dfrac{1}{\tau_{g,b}} + \dfrac{1}{\tau_i} + \dfrac{1}{\tau_p} + \Delta\left(\dfrac{1}{\tau_d}\right) + \Delta\left(\dfrac{1}{\tau_{p,d}}\right)$. The first three terms on the right are contributed to the unique scatterings (grain boundary scattering, interface region scattering, and precipitate scattering) in the nanocrystalline bulks. The last two are caused by the dislocation scattering and point defect scattering because of higher concentration. For the phonons with long-wavelength comparable to the grain size, the grain boundaries scatter the long-wavelength phonons independently of frequency, with a boundary scattering rate $\tau_{g,b}^{-1} = v/D_g$, where v is the group velocity of phonons and D_g is the equivalent diameter of a grain ($D_g \sim 10nm - 3\mu m$). The nano-precipitates in grains scatter middle-wavelength phonons, especially phonons with wavelength comparable to the precipitate size. Similar to grain boundary scattering, the phonon scattering rate of the precipitates $\tau_p^{-1} = v/D_p$ (D_p is the equivalent diameter of a precipitate, 3 nm - 100 nm). The linear dimensions of the point defects are much smaller than the phonon wavelength and the point-defect scattering rate $\tau_{p,d}^{-1} \propto N_{p,d}/(v^3)$, where $N_{p,d}$ is the fractional concentration of point defects. Compared with the ingots, the point defects will enhance the point-defect scattering rate of phonons $\tau_{p,d}^{-1}$ 100 - 1000 times, decreasing lattice thermal conductivity k_L efficiently.

We modeled the transport properties based on the Boltzmann equation and the relaxation time approximation, including the interfacial potential, and obtained the lattice contributions to the thermal conductivity as shown in Figure 8. The modeled results show that phonon contributions to the lattice thermal conductivity are reduced by a factor of two.

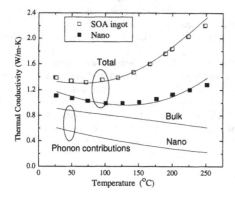

Figure 8. Thermal conductivity of $Bi_xSb_{2-x}Te_3$ nanocrystalline bulk alloy. Solid- and hollow-squares represent the experimental results for an SOA ingot and our nanocrystalline bulk alloys, respectively. Solid lines represent the corresponding calculations of the total and lattice part of the thermal conductivity, respectively.

Isotropic Thermal Properties and Temperature Stability

The hot-pressed nanocrystalline samples are expected to be isotropic because of the random orientation of the nanograins. To confirm the isotropic property in the hot-pressed nanocrystalline samples, disks and bars were cut along and perpendicular to the press direction and then measured. Although individual properties may differ 5% within the two directions, the final ZT values are isotropic. Such nearly isotropic characteristics are the result of the random orientation of the nanograins, and are superior to zone-melting made SOA Bi_2Te_3-based alloys which have layered structures and, consequently, anisotropic thermoelectric properties. To test the temperature stability of the nanocrystalline bulk samples, the same samples were repeatedly measured up to 250 °C and no significant property degradation was observed.

CONCLUSIONS

In summary, nanocrystalline bulks of p-type BiSbTe alloys were successfully made by hot-pressing nanopowders that were ball milled from either crystalline ingots or from elements. The nanocrystalline bulks were of nanograins containing nanoprecipitates. The unique nanostructures of the bulks reduced the thermal conductivity by increasing phonon scattering from imperfections in the bulks. ZT up to 1.4 was achieved at 100 °C. The nanocrystalline bulks are isotropic and thermally stable at high temperature. Such a simple approach should be readily applied to other thermoelectric materials that are hard to be made into crystalline ingot.

ACKNOWLEDGMENTS

The work is supported by DOE DE-FG02-00ER45805 (ZFR), and DOE DE-FG02-02ER45977 (GC), and NSF NIRT 0506830 (GC and ZFR).

REFERENCES

1. D. M. Rowe, Ed., *CRC Handbook of Thermoelectrics* (CRC Press, Boca Raton, FL, 1995).
2. H. J. Goldsmid, *Thermoelectric Refrigeration* (Plenum Press, New York, 1964).
3. T. M. Tritt, Ed., *Semiconductors and Semimetals* (Academic Press, San Diego, CA, 2001).
4. F. J. Disalvo, *Science* 285, 703 (1999).
5. B. C. Sales, *Science* 295, 1248 (2002).
6. R. Venkatasubramanian, E. Siivola, T. Colpitts, and B. O'Quinn, *Nature* **413**, 597 (2001).
7. T. C. Harman, P. J. Taylor, M. P. Walsh, and B. E. LaForge, *Science* **297**, 2229 (2002).
8. K. F. Hsu, S. Loo, F. Guo, W. Chen, J. S. Dyck, C. Uher, T. Hogan, E. K. Polychroniadis, and M. G. Kanatzidis, *Science* **303**, 818 (2004).
9. B. Poudel, Q. Hao, Y. Ma, Y. C. Lan, A. Minnich, B. Yu, X. Yan, D. Z. Wang, A. Muto, D. Vashaee, X. Y. Chen, J. M. Liu, M. Dresselhaus, G. Chen, and Z. F. Ren, *Science* **320**, 634 (2008).
10. Y. Ma, Q. Hao, B. Poudel, Y. C. Lan, B. Yu, D. Z. Wang, G. Chen, and Z. F. Ren, *Nano Letters* **8**, 2580 (2008).
11. Y. C. Lan, B. Poudel, Y. Ma, D. Z. Wang, M. S. Dresselhaus, G. Chen, and Z. F. Ren, *Nano Letters* (2009) (in press)
12. G. Joshi, H. Lee, Y. C. Lan, X. W. Wang, G. H. Zhu, D. Z. Wang, R.W. Gould, D. C. Cuff, M. Y. Tang, M. S. Dresselhaus, G. Chen, and Z. F. Ren, *Nano Letters* **8**, 2580 (2008).
13. X. W. Wang, H. Lee, Y. C. Lan, G. H. Zhu, G. Joshi, D. Z. Wang, J. Yang, A. J. Muto, M. Y. Tang, J. Klatsky, S. Song, M. S. Dresselhaus, G. Chen, and Z. F. Ren, *Appl. Phys. Lett.* **93**, 193121 (2008).
14. X. B. Zhao, X. H. Ji, Y. H. Zhang, T. J. Zhu, J. P. Tu, and X. B. Zhang, *Appl. Phys. Lett.* **86**, 062111 (2005).
15. X. F. Tang, W. J. Xie, H. Li, W. Y. Zhao, and Q. J. Zhang, *Appl. Phys. Lett.* **90**, 012102 (2007).
16. T. C. Harman, S. E. Miller, and H. L. Goeing, *Bull. Amer. Phys. Soc.* **30**, 35, (1955).
17. T. Thonhauser, G. S. Jeon, G. D. Mahan, and J. O. Sofo, *Phys. Rev. B.* **68**, 205207 (2003).

Mater. Res. Soc. Symp. Proc. Vol. 1166 © 2009 Materials Research Society   1166-N04-07

Thermal Transport in Rough Silicon Nanowires for Thermoelectric Applications

Sanjiv Sinha[1], Bair Budhaev[2] and Arun Majumdar[2,3,4]

[1]Mechanical Science & Engineering, University of Illinois, Urbana, IL 61801, USA
[2]Mechanical Engineering, University of California, Berkeley, CA 94720, USA
[3]Materials Sciences Division and [4]Environmental Energy Technologies Division, Lawrence Berkeley National Laboratory, Berkeley, CA 94720, USA

ABSTRACT

The coefficient of merit, ZT of nanostructured thermoelectric materials increases with reduction in thermal conductivity through phonon scattering. The ideal thermoelectric is considered to be an electron crystal and a phonon glass. This paper explores such a material concept by developing a theory for phonon localization in rough nanowires with crystalline cores. Results based on this theory suggest that the reported hundredfold decrease in thermal conductivity of rough silicon nanowires arises due to multiply scattered and localized phonons. Phonon localization presents a new direction to further enhance ZT through nanostructuring.

INTRODUCTION

Recently reported data on electrolessly etched rough silicon nanowires [1] suggest a hundredfold increase in ZT due to reduction in thermal conductivity of the same order. However, no known theoretical model for thermal conductivity of crystalline dielectrics predicts such reduction. Further, the observed linear increase in thermal conductivity with temperature in nanowires with diameters approaching 50 nm defies the widely-accepted Klemens-Callaway [2] model of thermal conductivity for silicon.

Here we discuss a new model that considers the wave-like transmission of acoustic phonons through the wire. We show that such a nanowire behaves in essence, like a waveguide with the resulting wave dispersion becoming size and roughness dependent. Considering a two-dimensional analogue of the wire problem, we solve the Helmholtz wave equation for acoustic waves in a rough film [3]. We show that roughness leads to a coupling of different modes and also the conversion of propagating modes to evanescent modes. This conversion to evanescence localizes phonons in wires with lengths approximately greater than a micrometer. We compute the resulting thermal conductivity of rough Si nanowires and show good comparison with experimental data for diameters less than 100 nm.

THEORY OF PHONON LOCALIZATION IN NANOWIRES

The objective of this paper is to understand the influence of a rough surface on phonon transport. A Boltzmann equation based transport is able to describe the reduction in thermal conductivity of smooth silicon nanowires [4] after accounting for the proper dispersion in such wires and boundary scattering along the lines of the Klemens-Callaway model. However, this approach fails to predict a nearly amorphous-like thermal conductivity in rough silicon nanowires. Instead of the typical Boltzmann picture, we adopt a wave description of phonon transport in this paper. The principal difference is that we need to account for phase in the wave description. The scattering of waves at rough boundaries may be modeled through the Rayleigh or the Helmholtz-Kirchoff approximation [5]. Since we are interested in considering multiply scattered phonons, we will adopt the Helmholtz-Kirchoff approach in this paper. In order to apply multiple scattering theory and yet keep the problem tractable, we make several concessions. First, we only consider a two-dimensional version of the problem. We will consider the actual cylindrical geometry in future work. Second, we describe phonon modes in a nanowire using a scalar wave equation. Rigorously, this description only applies to long-wavelength modes and ignores polarization. However, the scalar wave equation keeps the problem tractable and reveals the physics of localization clearly. In using a scalar wave equation, we follow previous work by Lifshitz and Cross [6] who computed the thermal conductance of silicon nitride bridges at low temperatures. Third, we use Dirichlet boundary conditions instead of the correct Neumann boundary conditions. This is an issue of mathematical convenience and will be corrected in future work. We note that the main difference between the two is the presence of the zeroth order mode in the Neumann problem. However, we do not expect the attenuation coefficient to vary widely between the two problems.

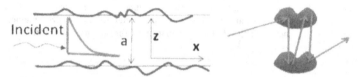

Fig. 1(a): We consider a two-dimensional analogue of the nanowire as shown. A phonon entering the wire undergoes boundary scattering from the rough surfaces. The mean wave function attenuates into the wire due to scattering from coherent to incoherent parts. (b) A depiction of multiple scattering of phonons between two scatterers. Consideration of such multiple scattering is key to predicting the reduced conductivity.

Removing the standard time dependence of $e^{-i\omega t}$ from the wave equation gives the Helmholtz equation, which is the starting point for this analysis,

$$\Delta \psi + K^2 \psi = 0 \tag{1}$$

where ψ is the spatial wave function for the phonon. Phonons confined to propagate inside the nanowire obey the Neumann boundary condition corresponding to rigid walls. However, purely for mathematical convenience, we consider instead Dirichlet boundary conditions corresponding to soft walls. We seek to improve upon this in future work. Our subsequent approach follows the development in Ref. [3] which considers the problem of rough waveguides using a perturbation method applied to the Helmholtz-Kirchoff integral.

The method essentially treats the rough surface, Σ as a random perturbation of a smooth surface, S. We can construct the Green function for Σ, knowing the Green function for S. The structure of the Green function corresponds to multiple scattering of a phonon from the two boundaries of the nanowire. We consider phonons confined to travel between two boundaries at $z=\zeta(x,y)$ and $z=a+\zeta(x,y)$, where ζ is a random variable depicting roughness of the surface and a is the thickness of the waveguide and corresponds to the diameter of the original nanowire. We further choose Gaussian roughness which has been used elsewhere [3, 6]. The statistical properties of the surface are as follow: $\langle \zeta \rangle = 0$, $\langle \zeta^2 \rangle = \sigma$, $\langle \zeta(r_1)\zeta(r_2) \rangle = \sigma^2 W(|r_1 - r_2|)$. The Green function for the Helmholtz equation in (1) can be written as

$$(\Delta + K^2)G(R, R_0) = -4\pi\delta(R - R_0) \qquad (2)$$

where R and R_0 are the observation and source co-ordinates respectively for the Green function. The boundary conditions are:

$$G(R, R_0)\big|_{R \in \Sigma} = 0 \qquad (3)$$

A Taylor expansion of the boundary condition gives

$$G\big|_s + \zeta \frac{\partial G}{\partial z}\bigg|_s = 0 \qquad (4)$$

Using Green's theorem, we express the perturbed Green function, G in terms of the unperturbed function G_0 as

$$G(R, R_0) = G_0(R, R_0) - \frac{1}{4\pi} \int_s \frac{\partial G_0(R, r)}{\partial z} \zeta(r) \frac{\partial G(r, R_0)}{\partial z} dS \qquad (5)$$

Iterating and averaging gives the Dyson equation

$$\langle G \rangle = G_0 + \int G_0 \hat{Q} \langle G \rangle dS \qquad (6)$$

Where the operator $\hat{Q}$ represents multiple scattering from the boundaries.

PHONON ATTENUATION COEFFICIENT

We solve for modified phonon dispersion in a rough nanowire in this section. The net modification is that the wave vector for every mode in the smooth wire becomes complex in a rough wire with a small but non-zero imaginary component. Thus the mean field intensity, corresponding to the coherent component, is continuously attenuated as the wave propagates inside the wire due to multiple scattering into incoherent components. The attenuation coefficient can be obtained from the dispersion relation and is inversely related to the localization length of a particular mode. On averaging Eq. (6), odd moments in the operator $\hat{Q}$ vanish. The resulting infinite series is conveniently expressed as a Feynman diagrammatic sum. We refer the reader to Ref. [3] for further details on the summation. We proceed directly to the result here. The original wave vector along the x direction associated with the nth mode, k_n is modified as follows.

$$k_n \rightarrow k_n + \delta k_n$$

$$\text{Re}(\delta k_n) = \frac{q_n^2 \sigma^2}{\pi dk_n} \int_{-\infty}^{+\infty} q \cot(qd)\tilde{W}(k_n - k)dk \qquad (7)$$

$$\text{Im}(\delta k_n) = \frac{\sigma^2}{d^2} \sum_{m=1}^{N} \frac{q_n^2 q_m^2}{k_n k_m} \left[\tilde{W}(k_n - k_m) + \tilde{W}(k_n + k_m) \right]$$

where q_n is the transverse component of the wave vector along the z direction, and $\tilde{W}$ is the Fourier transform of W. the attenuation coefficient is equal to the imaginary shift as given by Eq. (7c).

97

RESULTS

Figure 2(a) plots the phonon localization lengths in a silicon nanowire based on Eq. (7c) as a function of wire diameter and temperature. We assume the root mean square roughness and the

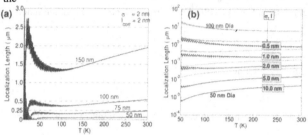

roughness correlation lengths to be each equal to 2 nm. This is a nominal value based on available SEM micrographs of rough nanowires [1]. Our results indicate that the localization length is less than 2 µm in these wires which is comparable to the lengths of the

Fig. 2 (a): Phonon localization lengths as a function of wire diameter and temperature. (b) Dependence of localization length on roughness for 50 nm and 100 nm diameter.

measured wires. Figure 2(b) shows the sensitivity of the localization length to the roughness. We find that the length scales nearly as the cubic power of roughness at 300 K assuming the correlation length and the rms roughness to be identical.

Figure 3(a) and (b) compare predictions of thermal conductivity as a function of temperature and wire diameter for different nanowires with diameters less than 100 nm. We express the thermal conductivity in terms of a Landauer formula as described in Ref. [6]. The multiple scattering theory described here is able to predict data for wires less than 100 nm reasonably well. However, it severely underestimates the conductance of wider wires. This is not surprising given the assumption of two-dimensions used in the analysis. Further, other scattering mechanisms common to Boltzmann transport have been ignored thus far. We have described only the first steps in constructing a theory of phonon transport including localization. A future analysis will include the correct boundary condition, the correct dimensionality and geometry and other scattering mechanisms in addition to localization.

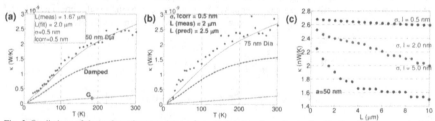

Fig. 3: Predictions of thermal conductance in rough nanowires based on phonon localization com compared with data from Ref. [1] for (a) 50 nm diameter and (b) 75 nm diameter wires. (c) Prediction of conductance for 50 nm wire for different roughness as a function of length at 300 K.

However, it is remarkable that despite all simplifying assumptions, the model is able to predict the thermal conductance of the thinnest nanowires over a wide temperature range. Figure

3(c) shows the thermal conductance of a 50 nm wire for different roughness at room temperature. Our model predicts that it is possible to further reduce conductance to approximately 1.5 nW/K. This requires a wire of 10 μm length and 5 nm surface roughness. This would represent a further doubling of ZT at room temperature.

This work develops a new model for phonon localization in rough silicon nanowires and provides quantitative guidance on further improving the ZT in such rough nanowires. A phonon localization model based on multiple scattering theory agrees reasonably with experimental data from the literature. Our work suggests that it is possible to further enhance ZT by exploiting the concept of a phonon glass – electron crystal.

ACKNOWLEDGMENTS

SS acknowledges Intel Corporation for supporting this post-doctoral work. He wishes to thank Renkun Chen and Kedar Hippalgaonkar for sharing data on nanowire measurements.

REFERENCES

1. A.I. Hochbaum et al., Nature 451, 163-167 (2008).
2. G. A. Slack, Solid State Physics, 34, 1, Academic Press, New York (1969).
3. V.D. Freilikher et al., Radiofizika, 13, 73 (1972).
4. Mingo, N. Phys. Rev. B, 68, 113308 (2003).
5. P. Beckmann and A. Spizzichino, *Scattering of Electromagnetic Waves from Rough Surfaces.* New York, NY: Artech House Publishers (1987).
6. D.H. Santamore and M.C. Cross, Phys. Rev. B, 63, 184306 (2001).

Theoretical Investigations

Mater. Res. Soc. Symp. Proc. Vol. 1166 © 2009 Materials Research Society   1166-N05-01

First Principles Study of Metal/Bi₂Te₃ Interfaces: Implications to Reduce Contact Resistance

Ka Xiong[a], Weichao Wang[a], Husam N. Alshareef[a,c], Rahul P. Gupta[a], John B. White[d], Bruce E. Gnade[a] and Kyeongjae Cho[a,b,*]

[a] Materials Science & Engineering Dept, The University of Texas at Dallas, Richardson, TX 75080, USA
[b] Physics Dept, The University of Texas at Dallas, Richardson, TX 75080, USA
[c] Materials Science & Engineering Dept, King Abdullha University of Science and Technology, Thuwal, Saudi Arabia
[d] Marlow Industries, 10451 Vista Park Road, Dallas, TX 75238, USA
*kjcho@utdallas.edu

ABSTRACT
We investigate the band offsets and stability for Ni/Bi₂Te₃ and Co/Bi₂Te₃ interfaces by first principles calculations. It is found that the surface termination strongly affects the band offsets. Ni and Co are found to form Ohmic contacts to Bi₂Te₃. The interface formation energies for Co/Bi₂Te₃ interfaces are much lower than those of Ni/Bi₂Te₃ interfaces. Our calculations are consistent with the experimental data.

INTRODUCTION

The solid-state thermoelectric (TE) cooler has been of great interest in many applications because of its advantages of reliability, low-noise operation, miniaturization and high power density[1-3]. Commercial TE cooling devices use doped Bi₂Te₃ ((Bi,Sb)₂Te₃ for p-type and Bi₂(Te,Se)₃ for n-type) as the TE elements due to its high figure of merit (ZT ~1)[4] at room temperature. To allow TE coolers to reach the next level of performance in terms of efficiency and power density, the size of the device needs to be scaled down. Consequently, the electrical resistance between the contact metal and the TE material plays an important role in the device performance[5], as the material "figure of merit" (Z) is degraded by the contact resistance and the relationship between Z and Z_D (device "figure of merit") can be shown in the following equation

$$Z_D = \left(\frac{L}{L + 2r_c\sigma} \right) Z \quad (1),$$

where L is the device length, r_c is the contact resistance and σ is the bulk conductivity. The resulted maximum coefficient of performance (COP) of the device can be expressed as[6]

$$COP = \frac{Q_C}{W} = \frac{T_C}{T_H - T_C} \times \frac{\sqrt{1 + Z_D T} - \frac{T_H}{T_C}}{\sqrt{1 + Z_D T} + 1} \quad (2),$$

where Q_C is the heat pumping capacity, W is the input power. T_C and T_H represent the temperature of the cold side and the hot side, respectively. T is the average temperature. Therefore, from a device point of view, although a high Z of the material has been achieved, the COP can still be low due to the degradation of Z by the contact resistance etc. So merely seeking how to improve Z is not adequate. For current bulk TE devices, electroless Ni has been used as the contact metal with a contact resistance (r_c) of ~ 5x10⁻⁶ Ω cm². The contact resistance needs to be at least 10-100 times lower in order to maintain the required device scaling[7]. To see why the contact resistance is critical, we can do a simple quantitative analysis on how the contact resistance influences the device dimensions and hence the COP. The electrical conductivity (σ) of Bi₂Te₃ at room temperature is of order 1×10^3 (Ω•cm)⁻¹. By substituting r_c and σ into eqn (1) we can plot the Z_D/Z as a function of device length L, as shown in Fig. 1a. It can be seen that for L < 100 μm, the contact resistance significantly influences Z_D/Z. If we require Z_D equal to 90% of Z, then from the plot we find that the device length (L) is of ~900 μm for r_c =5x10^{-6} Ωcm^2. If r_c is 10 times lower (5x10^{-7}), L would be ~10 times smaller, which is of 100 μm. The corresponding COP is degraded by ~10%, as shown in Fig. 1b. Thus, it is necessary to lower r_c by engineering the interface between the contact metal and TE materials.

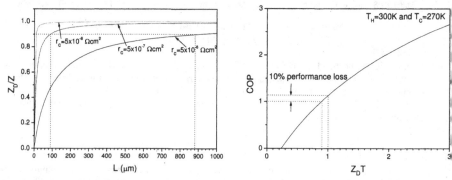

Fig.1 (a) The ratio of device figure of merit Z_D and material figure of merit Z versus the device length. (b) Coefficient of performance (COP) at room temperature as a function Z_DT.

The contact resistance may arise from the formation of the Schottky barrier between the metal and the semiconductor, or the interface band alignment has not been systematically optimized. Therefore, it is necessary to understand the factors that control the Schottky barrier height between metal and thermoelectric material.

Recently, we have investigated experimentally the Ni/Bi$_2$Te$_3$ and Co/Bi$_2$Te$_3$ interfaces. Full details of the experiment will be published elsewhere. Ni and Co films were sputtered on polycrystalline bulk Bi$_2$Te$_3$ (Se-doped) with thicknesses of 120 nm and 90 nm, respectively. After post-annealing at 200°C, for the Ni/Bi$_2$Te$_3$ interface a 460 nm NiTe interfacial region has been observed[8], while for the Co on Bi$_2$Te$_3$ the interface is rather sharp, with a very thin interfacial region (~20 nm). The formation of thick NiTe interfacial region is consistent with previous experimental work[9]. For the purpose of understanding detailed interface electronic structures, we use first principles calculations to investigate the interface chemistry and stability of Ni/Bi$_2$Te$_3$ and Co/Bi$_2$Te$_3$ interfaces[10].

COMPUTATIONAL METHODS

We use density-functional calculations to estimate the band offsets and formation energy of several Ni/Bi$_2$Te$_3$ and Co/Bi$_2$Te$_3$ interfaces. We employ the total energy plane-wave basis code VASP[11]. The exchange correlation energy is approximated by the generalized gradient approximation (GGA). The pseudopotential is generated by projected augmented wave (PAW) method. A 6x6x1 k-point Monkhorst-Pack grid was sufficient for calculating total energies and a 10x10x1 grid for the density of states (DOS). Since the electronic structure of Bi$_2$Te$_3$ is strongly affected by spin-orbit coupling (SOC), as reported by many workers, we carry out our interface calculations by including SOC.

The interfaces are modeled by a superlattice containing one interface and 10Å vacuum on top. To choose the metal surface for interface model, we can either use the experimental observation by constructing a large lattice-match superlattice, or find a suitable orientation that could give a reasonable cell size. The latter method has been adopted for a compromise between the computational cost and accuracy. We consider the Ni(111) / Bi$_2$Te$_3$(0001) junction, which resulting a lattice mismatch of 1%. The Ni(111) slab is ~10Å thick with 4 layers of Ni, while the Bi$_2$Te$_3$(0001) slab is ~40Å thick with 8 layers of Bi and 11-13 layers of Te. This supercell is sufficient to converge the calculated band offsets. We consider two types of interfaces: Bi-terminated and Te-terminated (Fig.2a,b). We found the work functions of these metals are insensitive to the applied strain. In our model, we use pure Bi$_2$Te$_3$, as in reality Se doping primarily lowers the thermal conductivity of Bi$_2$Te$_3$ rather than modifies the electronic structure of Bi$_2$Te$_3$. Moreover, the experimental data show that the interfacial region does not contain Se. In each case the atomic positions are relaxed and the in-plane lattice constants and all angles are kept fixed.

To determine the relative stability of these metal/Bi$_2$Te$_3$ interfaces, we calculated their interface formation energies. This is a function of chemical potentials and can be written as[12]

$$E_{form}^{metal/Bi_2Te_3} = E_{total}^{metal/Bi_2Te_3} - n_{metal}E_{metal} - \frac{n_{Bi}}{2}E_{Bi_2Te_3} - (n_{Te} - \frac{3}{2}n_{Bi})\mu_{Te} \quad (3)$$

where $E_{total}^{metal/Bi_2Te_3}$ is the total energy of the given supercell, n_x is the number of atoms of element x in it; $E_{Bi_2Te_3}$ and E_{Metal} are the total energies per formula unit in Bi$_2$Te$_3$ and the metal (Ni or Co), respectively. There is a single independent parameter determined by the growth conditions, the Te chemical potential μ_{Te}. Its highest value is μ_{Te}^{bulk}, the energy of bulk Te chemical potential. Its lowest value is for thermodynamic equilibrium with Bi$_2$Te$_3$. Taking the formation enthalpy of bulk Bi$_2$Te$_3$ as $\Delta H \sim -1.3$ eV[13], the limiting values of μ_{Te} are $\mu_{Te} = \mu_{Te}^{bulk} + \frac{\Delta H}{3}$ (Bi-rich, favoring Te substitutions) and $\mu_{Te} = \mu_{Te}^{bulk}$ (Te-rich, favoring Bi substitutions).

RESULTS

Fig.3 shows the calculated band structure of bulk Bi$_2$Te$_3$. Without considering the SOC, we obtained a direct band gap of 0.4 eV, with both conduction band minimum and valence band maximum lying at Γ. The SOC effect scales the band gap down to 0.13 eV, close to the experimental band gap of 0.16 eV[14], in good agreement with other theoretical results[15-19]. The CBM and VBM now locate at a point between Z and F. It is noted that along the Bi$_2$Te$_3$[0001] (Γ to Z), the SOC effect reduces the band dispersion. The DOS plot (not shown) shows that the conduction band consists of Bi 6p states while the valence band consists of Te 5p states.

The relaxed structures (Fig.2) for Bi-terminated and Te-terminated Ni/Bi$_2$Te$_3$ interfaces shows that the interfaces are formed by either Bi-Ni or Te-Ni bonds. The interfacial Bi and Te atoms are 6-fold coordinated, as in bulk Bi$_2$Te$_3$. Each interfacial Bi (or Te) atom bonds to 3 Ni atoms. The interfacial Ni atoms are 9- or 10-fold coordinated. The calculated Bi-Ni and Te-Ni bond lengths range from 2.4 to 2.6 Å. Co/Bi$_2$Te$_3$ interfaces have similar interfacial bonding because Co has similar crystal structure and lattice constant to Ni.

Fig.4 shows the interface formation energy for the various interfaces as a function of the tellurium chemical potential μ_{Te}. We can see that the formation energy of the Bi-terminated interface increases with increasing μ_{Te}, while the energy of the Te-terminated interface is independent with μ_{Te}. The Bi-terminated interface is always more stable than the Te-terminated interface. More importantly, the formation energies for Co/Bi$_2$Te$_3$ interfaces are much lower than Ni/Bi$_2$Te$_3$ interfaces, by ~7 eV! This result may explain why Ni interacts more readily with Bi$_2$Te$_3$ to form a thick NiTe interfacial region while Co does not. The formation for the Ni/Bi$_2$Te$_3$ interface is less energetically expensive so that Ni tends to diffuse into Bi$_2$Te$_3$ to form an interface with lower energy.

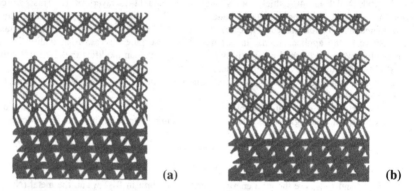

Fig.2 The interface region of the relaxed atomic structures for (a) Bi-terminated and (b) Te-terminated Ni(Co)/Bi$_2$Te$_3$ interfaces. Ni(Co) in light blue, Te in brown, and Bi in purple.

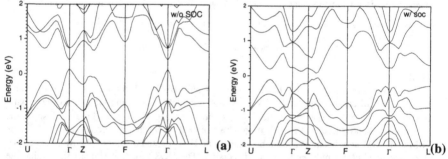

Fig.3 Calculated band structure and density of states of bulk Bi$_2$Te$_3$. (a) band structure without considering SOC, (b) band structure including SOC.

The band offset is derived from the local DOS, as shown in Fig.5. The plots show that DOS on atoms well away from the interface replicate those of bulk atoms, as there are no metal states in the Bi$_2$Te$_3$ band gap, while the DOS of interfacial atoms show the expected tailing of metal states into the Bi$_2$Te$_3$ band gap. The band offsets are then calculated as the energy difference of the bulk Bi$_2$Te$_3$ valence band edge to the bulk Ni Fermi level (denoted as VBO). From the DOS we can see that for Bi-terminated Ni/Bi$_2$Te$_3$ interface, the Fermi level lies at the valence band edge so that this interface forms a p-type Ohmic contact. In contrast, for Te-

terminated Ni/Bi₂Te₃ interface, the Fermi level now lies at the conduction band edge, which gives an n-type Ohmic contact. Thus, the band offset depends on the interfacial bonding.

The formation of different types of Ohmic contact with respect to the termination of the interfaces can be explained in terms of local bonding. The formation of the metal/semiconductor interface causes charge transfer from the metal to the semiconductor, due to their different electronegativity (EN). Despite the small differences of EN values (Pauling criterion) of the elements in our system (Ni: 1.9, Co: 1.88, Bi: 2.02, Te: 2.1), the small band gap of Bi₂Te₃ makes the Fermi level easily swipe across the band gap. Therefore, we conclude that a pure Ni/Bi₂Te₃ or Co/Bi₂Te₃ interface can form a good Ohmic contact. The sources for degrading the Ohmic contact must be extrinsic. This indicates that Co and Ni are suitable contact materials. Co is a better choice because of its low diffusivity.

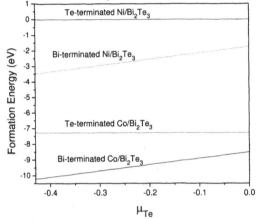

Fig.4 Interface formation energies of various interfaces vs Te chemical potential.

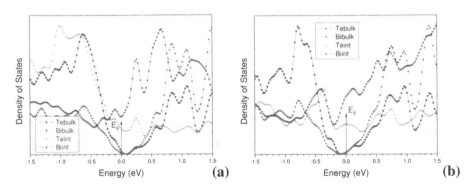

Fig.5 Partial density of states (DOS) for (a) Bi-terminated and (b) Te-terminated Ni/Bi₂Te₃ interfaces. Tebulk and Bibulk denote the Te and Bi atoms away from the interface. Teint and Biint denote the Te and Bi atoms at

CONCLUSIONS

In summary, first principles calculations of band offsets and stability of Ni/Bi_2Te_3 and Co/Bi_2Te_3 interfaces are presented. We found that the Bi-terminated interface always has lower formation energy than the Te-terminated interface. The formation energies of Co/Bi_2Te_3 interfaces are much lower than Ni/Bi_2Te_3 interfaces, making the Ni/Bi_2Te_3 interface more prone to reaction as temperature is increased. Our results are consistent with experimental data. All the interfaces give Ohmic contact. Therefore, we conclude that the source for degrading the Ohmic contact must be extrinsic. This suggests that processes should be optimized to avoid introducing impurities, defects or oxidation into the metal/Bi_2Te_3 contact.

ACKNOWLEDGMENTS

This research is supported by II-VI foundation, a private foundation. We thank Dr. Jeff Sharp for useful discussions. Calculations were performed on the clusters in TEXAS ADVANCED COMPUTER CENTER (TACC).

REFERENCES

1. M. S. Dresselhaus, G. Chen, M. Y. Tang, R. Yang, Z. Ren, Adv. Mater. 19, 1043, (2007).
2. G. J. Snyder and E. S Toberer, Nat. Mater. 7, 105, (2008).
3. V. Semenyuk, Proc 22nd Intl. Conf. on Thermoelectrics, pp631, Montpellier, France (2003).
4. G. S. Nolas, J. Sharp and H. J. Goldsmid, "Thermoelectrics – Basic Principles and New Materials Developments", Springer (2001).
5. J. -P. Fleurial, Proc. 18th Intl. Conf. on Thermoelectrics, pp294, Baltimore, USA (1999).
6. L.I. Anatychuk, Proc. 15th Intl. Conf. on Thermoelectrics, pp 279, Pasadena, USA(1996)
7. V. Semenyuk, Proc 20th Intl. Conf. on Thermoelectrics, pp391, Beijing, CHINA (2001).
8. O. D. Iyore, T.H. Lee, R. P. Gupta, J. B. White, H. N. Alshareef, M. J. Kim and B. E. Gnade, Surf. Interface Anal. (2009), in press.
9. Y. C. Lan, D. Z. Wang, G. Chen and Z. F. Ren, Appl. Phys. Lett. 92, 101910 (2008).
10. We have used first principles method to study similar metal/oxide interface band offset problems: B. Magyari-Kope, S. Park, L. Colombo, Y, Nishi, and K. Cho, J. Appl. Phys. 105, 013711 (2009).
11. G. Kresse and J. Furthmuller, Comput. Mater. Sci. 6, 15 (1996); Phys. Rev. B 54, 11169 (1996).
12. K. Xiong, P. Delugas, J. Hooker, V. Fiorentini, J. Robertson, Appl. Phys. Lett. 92, 113504 (2008).
13. N. P. Gorbachuk, and V. R. Sidorko, Powder Metallur. Metal Ceram., 43, 284 (2004).
14. H. J. Goldsmid, Thermoelectric Refrigeration (Plenum, New York, 1964).
15. B. –L. Huang and M. Kaviany, Phys. Rev. B 77, 125209 (2008).
16. T. J. Scheidemantel, C. Ambrosch-Draxl, T. Thonhauser, J. V. Badding, and J. O. Sofo, Phys. Rev. B 68, 125210 (2003).
17. S. J. Youn and A. J. Freeman, Phys. Rev. B 63, 085112 (2001).
18. P. Larson, Phys. Rev. B 68, 155121 (2003).
19. M. Kim, A. J. Freeman, and C. B. Geller, Phys. Rev. B 72, 035205 (2005).

Electronic Structure and Transport Properties of Ternary Skutterudite: $CoX_{3/2}Y_{3/2}$

Dmitri Volja[1], Marco Fornari[2,] Boris Kozinsky[3] and Nicola Marzari[1]
[1] Massachusetts Institute of Technology, Department of Materials Science and Engineering, Cambridge, MA 02139
[2] Central Michigan University, Department of Physics, Mount Pleasant, MI 48859
[3] Robert Bosch LLC, Research and Technology Center, Cambridge, MA 02142

ABSTRACT

Electronic properties of ternary skutterudites $AX_{3/2}Y_{3/2}$ (A=Co, X=Ge, Sn and Y=S, Te) are investigated using first principles calculations to clarify recent experimental results. Band derivatives are computed accurately within an approach based on Maximally Localized Wannier Functions (MLWFs). Band structures exhibit larger effective masses compared to parental binary $CoSb_3$. Our results also indicate a more parabolic dispersion near the top of the valence band and a multivalley character in both conduction and valence band. Despite the improved thermopower these skutterudites has relatively low power factor due to increased resistivity. The fundamental cause of such large resistivity seems to be associated with the ionicity of the bonding.

Figure 1. Distorted ReO₃ Figure FIG. 1: FIG

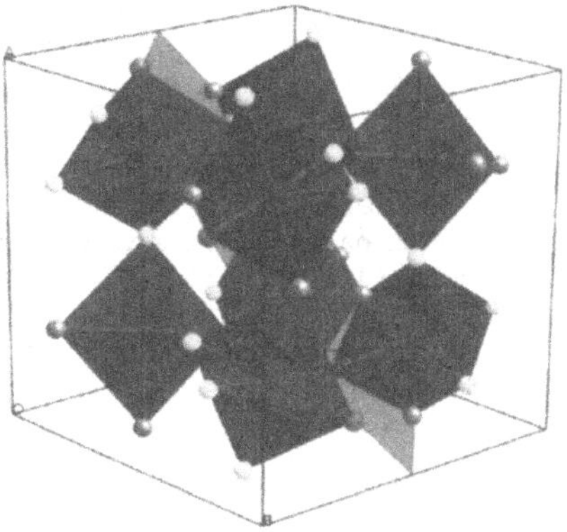

INTRODUCTION

Ternary unfilled skutterudites, derived from $CoSb_3$ by substituting group IV and group VI elements on the pnictogen site, were experimentally studied [1-5] because of their potential interest for thermoelectric applications at room temperatures and above. The stochiometric compounds are ordered within a double skutterudite cell with a small rhombohedral distortion (space group R-3) [2]. The structure can be derived from ReO₃ structure (heavily distorted by $a^+a^+a^+$ octahedral rotations) by substitution of O layers perpendicular to the [111] direction with alternating group IV and group VI layers. The rectangular rings formed with IV-VI elements are distorted. Two cages in the unit cell remain available for filling with large ions.

We studied $CoGe_{3/2}S_{3/2}$, $CoGe_{3/2}Te_{3/2}$, and $CoSn_{3/2}Te_{3/2}$ to investigate the effect of ion size and different electronegativity. Experimental results for these compounds [3-4] point to low thermal conductivity, large thermopower, but large resistivity. Experimentally $CoGe_{3/2}S_{3/2}$ has positive, p-type conduction behaviour with Seebeck coefficient of approximately 200 μV/K. Tellurium containing samples exhibit negative, n-type conductivity with Seebeck coefficient values of -710μV/K and -300 μV/K at room temperature for $CoGe_{3/2}Te_{3/2}$ and $CoSn_{3/2}Te_{3/2}$, respectively. Thermal conductivities at 300K are 2.0 W/m/K, 1.5 W/m/K and 0.6 W/m/K for $CoGe_{3/2}S_{3/2}$, $CoGe_{3/2}Te_{3/2}$ and $CoSn_{3/2}Te_{3/2}$, respectively [5].

METHODOLOGY

Our calculations were performed using density functional theory, in local density approximation (LDA). [6-9] We used plane waves and first principles pseudopotentials techniques as implemented in the Quantum-Espresso package. [10] The crystal symmetry of ternary systems is lowered from body-centered Im-3 (16 atoms) to rhombohedral R-3 (32 atoms). We used 24 Ry energy cutoff for the plane wave basis. The reciprocal space was sampled with a mesh of 64 k-points for self-consistent calculation. We fully relaxed the geometrical structure by minimizing the total energy at constant experimental volume and computed band structures and density of states. For comparison we have calculated within the same approach the properties of $CoSb_3$.

The evaluation of electronic transport properties requires integrating of functions in the reciprocal space over the Brillouin zone. The integral must be treated carefully in order to avoid inaccuracy due to possible band crossing. Extremely dense k-point meshes might be needed to achieve convergence. In this work we avoid the sampling issue by using a Maximally Localized Wannier Functions (MLWF) scheme for interpolating both the energy bands and the matrix elements of other periodic operators such has electron velocity. [11-13] In this approach, the first-principles calculation is performed on a relatively coarse 4x4x4 k-point grid. Then, in a post-processing step the calculated electronic structure is mapped onto an "exact" tight-binding Slater-Koster type model within the MLWFs basis. Not only the bands are interpolated without the loss of the accuracy, but also their k-space derivative can be evaluated analytically. By working in this basis the relevant transport quantities are evaluated inexpensively.

The phase arbitrariness in Wannier states construction is controlled by maximal localization criteria. As initial guess for the localization procedure of the Wannier Functions for the conduction manifold we place s-states on cation-anion bonds quarter away from the cation (Co) site. For the valence manifold basis construction as initial guess we use results from faster-converging zone center calculations. Calculation of band derivatives and transport coefficient has been implemented in the Wannier90 package. [14]

An interesting by-product of our methodology are the MLWFs that may indicate chemical features of the bonding.

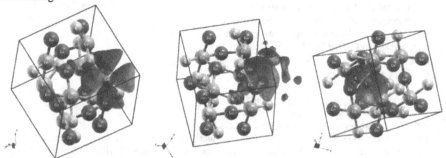

Figure 2. Maximally Localized Wannier Functions for valence manifold of $CoGe_{3/2}S_{3/2}$. Strong localization is seen. In addition one can recognize the atomic t_{2g} states associated with Co site and p-states bonding character between cation and anion sites.

STRUCTURAL FEATURES OF TERNARY SKUTTERUDITES

Minor structural differences were observed with respect to the experimental structure. In particular, we found off-centering of Co in the octahedral and distortions associated with four-membered rings. Most of these effects are justified on the basis of different anion sizes and ionicity

X,Y	Co - X	Co - Y	X - Y	X - Y	θ
Sb		2.85	2.98		
Ge, S	2.3	2.2	2.43	2.60	88
Ge, Te	2.3	2.5	2.76	2.93	87
Sn, Te	2.5	2.5	2.96	3.13	88

Table 1: Bonds lengths after relaxation with R-3 symmetry (Angstrom units). The angle indicates the deformation of four-membered rings.

ELECTRONIC STRUCTURE

The density of states clearly point to the ionicity of the X-Y bonding. Indeed, an energy gap appears in the anion-s manifold that follows the bond ionicity trends. In all three cases the majority of states in valence manifold right below the fermi level are Co-d states. The different contribution of anion-p orbitals in the valence manifold, also, reflect an increased ionic character of the bonding. The bottom of the conduction band remains mainly Co-d in character.

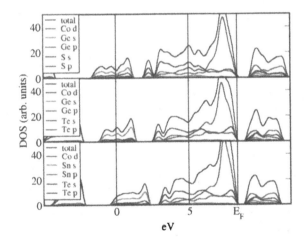

Figure 3. Total and partial density of states for $CoGe_{3/2}S_{3/2}$ (top) $CoGe_{3/2}Te_{3/2}$ (middle) and $CoSn_{3/2}Te_{3/2}$ (bottom)

All three compounds, we analyzed, are semiconductors with a direct band gap at Brillouin zone center that is significantly larger than in $CoSb_3$. While in the parent material the LDA band gap is 0.23 eV (under estimated by about 30% with respect to the experimental value), [15-16] for the ternaries we computed 0.61, 0.51 and 0.48 eV for $CoGe_{3/2}S_{3/2}$, $CoGe_{3/2}Te_{3/2}$ and $CoSn_{3/2}Te_{3/2}$ respectively.

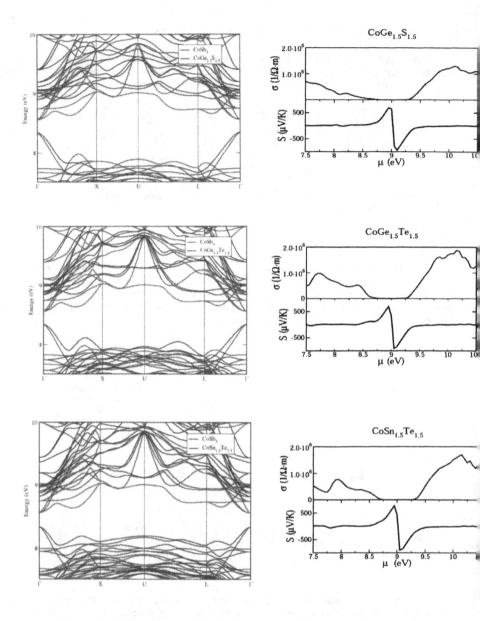

Figure 4: Band structures and electronic transport coefficients for three ternary skutterudites. S and σ are computed in the constant scattering time approximation assuming τ = 10 fs.

Significant changes also happen to the top valence band. This band near Γ changes its linear behavior to become more parabolic. While still linear close to Γ, the slope is significantly smaller. The dependence of power factor σS^2 on carriers concentration is more doping dependent. [17] The top of the valence band is formed by (pp-π*) orbitals of the four-membered rings and is sensitive to the substitution of Sb with IV-VI couples. The dispersion is decreased and, at Γ, the top of the valence band moves closer to the low lying heavy bands as the ionicity decreased from (Ge,S), (Ge,Te), to (Sn,Te). The lower lying heavy valence bands have multi-valley character with quasi-degenerate pockets. Large effective masses and multi-valley character are observed also in conduction band.

	m* (top VB at Γ)
$CoSb_3$	0.063
$CoGe_{3/2}S_{3/2}$	0.196
$CoGe_{3/2}Te_{3/2}$	0.169
$CoSn_{3/2}Te_{3/2}$	0.134

Table 2: CoSb3 has lowest effective mass among studied skutterudite systems. Caillat et.al. [15-16] ~0.071m$_e$

TRANSPORT COEFFICIENTS

In the Wannier basis all necessary ingredients for semiclassical Boltzman transport theory, such as group velocities, can be evaluated anayticaly without any loss of the accuracy. This is an improvement with respect to common finite differences approach. In addition, Wannier interpolation technique was used in order to avoid solving eigenvalue problem at each k-point of a dense k-point grid required for numerical integration. We show in Fig. 4 calculations at room temperature (300K). For the electrical conductivity we arbitrary fixed a constant scattering time at $\tau = 10$ fs and assumed rigid band approximation.

We computed the Seebeck coefficient values at optimal doping above 500 μV/K. This values are larger than what we found in equivalent calculations for CoSb$_3$ (~420 μV/K) and in reasonable agreement with the trend observed experimentally. [3-5] In the constant scattering time approximation the thermopower is not affected by the value of τ. The estimate of the electrical conductivity indicates that the band structure contribution to the conductivity is favorable. Since our calculations find favorable band structures and large S, the detrimental effect on the conductivity, found experimentally with respect to CoSb$_3$, can be attributed to a shorter scattering time. The cause of this may be the increased ionicity of the bonding.

CONCLUSIONS

We have investigated three unfilled ternary skutterudites that experimentally are reported to have significantly lower thermal conductivities. Within first-principles approach we analyzed their band structures and found significant changes in electronic states upon substitution with IV-VI couples with respect CoSb$_3$. The top valence band becomes more parabolic, the energy gap increases, and the both valence and conduction band exhibit multivalley character. The changes are mainly related to the ionicity of the bonds.

Using MLWF approach we interpolated accurately the band structure and its derivatives. We used the derivatives to compute transport coefficients in the BTE within the constant scattering time approximation. We observe the increase of S by ~20% compared to similar studies in $CoSb_3$. The increased resistivity is likely due to scattering mechanisms introduced by the ionic character of the substitution. Isovalent substitutions with large mass difference may overcome this problem.

ACKNOWLEDGMENTS

We thank D. Wee for useful discussions and Robert Bosch LLC for the financial support.

REFERENCES

1. G. S. Nolas, J. Yang, and R. W. Ertenberg, Phys. Rev. B 68, 193206 (2003)
2. F. Laufek, J. Návrátil and V. Goliáš, Powder diffraction 23, 15 (2008)
3. P. Vaquiero, G. G. Sobany and M. Stindl, J. of Solid State Chemistry 181, 768 (2008)
4. P. Vaqueiro, G. G. Sobany , A.V. Powell, and K. S. Knight, J. of Solid State Chemistry 179, 2047 (2006)
5. P. Vaquiero and G. G. Sobany, Mater. Res. Soc. Symp. Proc. Vol. 1044 (2008)
6. W. Kohn and L. J. Sham, Physical Review 140, A1133–A1138 (1965).
7. P. Hohenberg and W. Kohn, Physical Review 136, B864-B871 (1964).
8. D. M. Ceperley and B. J. Alder, Phys. Rev. Lett. 45, 566 (1980)
9. J. P. Perdew and A. Zunger, Phys. Rev. B 23, 5048 (1981).
10. S. Baroni, A. Dal Corso, S. de Gironcoli, P. Gianozzi, C. Cavazzoni, G. Ballabio, S. Scandolo, G. Chiarotti, P. Focher, A. Pasquarello, K. Laasonen, A. Trave, R. Car, N. Marzari, and A. Kokalj, http://www.pwscf.org/
11. N. Marzari and D. Vanderbilt, Phys. Rev. B 56, 12847 (1997)
12. J. R. Yates, X. Wang, D. Vanderbilt and I. Souza, Phys. Rev. B 75, 195121 (2007)
13. I. Souza, N. Marzari, and D. Vanderbilt, Phys. Rev. B 65, 035109 (2001).
14. A. A. Mostofi, J. R. Yates, Y.-S. Lee, I. Souza, D. Vanderbilt, and N. Marzari, Comput. Phys. Commun. 178, 685 (2008), http://www.wannier.org
15. J.-P. Fleurial, T. Caillat, A. Borshchevsky, Proceedings of the 16th International Conference on Thermoelectrics, Dresden, Germany, 1997, p. 1.
16. J.-P. Fleural, T. Caillat and A. Borschevsky, Proc. 14th Int. Energy Conf. on Thermoelectrics St. Petersburg, Russia,(1995), p.231
17. D.J. Singh and I.I. Mazin, PRB 56, R1650 (1997)

Mater. Res. Soc. Symp. Proc. Vol. 1166 © 2009 Materials Research Society   1166-N05-07

Direct and Indirect Effects of Filling on Electronic Structure of Skutterudites

Daehyun Wee[1], Boris Kozinsky[2,4], Marco Fornari[3], and Nicola Marzari[4]
[1]Robert Bosch LLC, Research and Technology Center, Palo Alto, CA 94304, U.S.A.
[2]Robert Bosch LLC, Research and Technology Center, Cambridge, MA 02142, U.S.A.
[3]Central Michigan University, Department of Physics, Mt. Pleasant, MI 48859, U.S.A.
[4]Massachusetts Institute of Technology, Department of Materials Science and Engineering, Cambridge, MA 02139, U.S.A.

ABSTRACT

We perform ab-initio computations to investigate the family of $CoSb_3$ skutterudites in an attempt to develop deeper understanding of the effect of fillers. Primary focus is on Ba-filled $CoSb_3$ systems, while Ca and Sr-filled systems are also compared for checking consistency. We analyze both global and local structural effects of filling. We show the specific deformation of Sb network introduced by the filler. Such a deformation is localized around the filler site since soft Sb rings accommodate the distortion. Rearrangement of Sb atoms affects the band structures, and we perform additional analysis to clarify the effect of volume on the band gap. Phonon dispersions are briefly discussed, and filler-dominated vibrations are identified. These modes form the first optical modes at Γ. They manifest themselves in phonon dispersion curves as flat lines, showing that they are localized, while filler vibration is strongly coupled with nearby Sb atoms.

INTRODUCTION

Filled skutterudites are a promising class of medium to high-temperature thermoelectric materials. They exhibit low thermal conductivity, which is typically attributed to the phonon scattering process introduced by the filler. In recent years, however, the exact nature of the physical process involved in the reduction of the thermal conductivity and the associated rattling motion of the filler atom have been the subject of intense investigations and controversy. To provide better insights on the issue, we analyze the effect of filling $CoSb_3$ with alkaline earth metals on the electronic band structure and on the vibrational spectra, using state-of-the-art first principles calculations.

METHODOLOGY

Calculations are performed using density functional theory (DFT) and density functional perturbation theory (DFPT) [1] with Quantum-Espresso software [2], with LDA functionals. Transport properties, including Seebeck coefficient, are studied by solving the Boltzmann transport equation under constant-relaxation-time approximation, using the BoltzTraP package [3]. The unfilled and fully filled (Ca, Sr, Ba) $CoSb_3$ compounds are considered using a body-centered cubic (BCC) unit cell with 16(+1) atoms. The compounds with uniform 50% filling are obtained from a simple cubic (SC) unit cell of 32+1 atoms. We obtain 25% and 75% Ba-filled structures from a tetragonal (TetP) 1x1x2 supercell. Finally, a 2x2x2 simple cubic cell with the filling fraction of 1/16 is used to investigate the structure of the filler-induced distortion. A kinetic energy cutoff of 30 Ry is used for all our electronic structure calculations. A 6x6x6 k-point grid is employed for body-centered cubic cells, and equivalent sampling is used for larger cells. For phonon calculations, the dynamical matrix is explicitly computed on a 2x2x2 grid by linear response DFPT calculations, whose result is later Fourier interpolated onto a finer mesh.

GLOBAL AND LOCAL STRUCTURAL CHANGES UNDER FILLING

As presented in Figure 1, the lattice size a_0 increases in a linear fashion, as the size and amount of the filler increases. Unfilled $CoSb_3$ lattice size in ground state is calculated at 8.972 Å, which is smaller than the experimental value 9.035(1) Å [4]. However, when thermal expansion is considered within quasi-harmonic approximation [5], the lattice size becomes larger: 9.017 Å at 300 K, which is closer to the experimental value. The locations of Sb atoms in unfilled $CoSb_3$ are fixed by two internal coordinate symmetry parameters y and z. Our values are close to experiment [6], as shown in Table 1.

We study the internal distortion introduced by the filler and its range of influence by analyzing the computationally relaxed atomic structure in a large supercell (SC 2x2x2) containing 256+1 atoms, or Ba filling fraction of 1/16. Closest Sb-Sb pairs can be classified into two distinct groups: each Sb atom belongs to an orthogonal ring consisting of four Sb atoms, coplanar with two nearest filler sites. The alignment of each pair can be either longitudinal along the direction connecting two nearest filler sites, or transverse to the direction. In unfilled $CoSb_3$, longitudinal pairs are longer than transverse ones. Filling, however, introduces local stress and rearranges Sb atoms, making transverse pair longer and longitudinal pairs shorter. This effect is demonstrated in Figure 2, where the lengths of pairs just beside a Ba atom are significantly different from those observed in unfilled $CoSb_3$. However, it is localized around filler sites. The next-nearest pairs barely see the presence of the filler, and the length of Sb-Sb pairs rapidly approaches the corresponding value in the unfilled $CoSb_3$. On the other hand, Co-Sb pairs are apparently much more rigid, and do not show as much variation due to the insertion of Ba atoms. This observation reveals that the structural effect of filling is rather localized around each filler site due to softness of Sb rings.

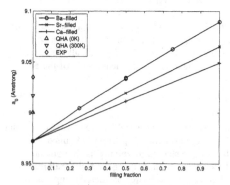

	Theory	EXP
a_0 (Å) 300K	9.017	9.035
y	0.3335	0.335 37
z	0.1589	0.157 88

Table 1. Lattice and internal coordinate parameters. Values are compared to experimental data [4,6].

Figure 1. Lattice a_0 versus filling fraction.

116

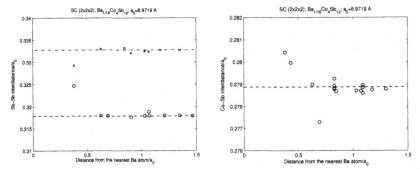

Figure 2. (Left) Length of closest Sb-Sb pairs versus the distance of the pair from its nearest Ba filler atom (in units of a_0). Left: circles are transverse pairs; crosses: longitudinal pairs; reference lines represent lengths in unfilled $CoSb_3$; dash-dot. Right: Length of closest Co-Sb pairs as a function of the distance from the filler site.

ELECTRONIC AND VIBRATIONAL SPECTRA

Unfilled $CoSb_3$ is a direct gap insulator with a moderate band gap of around 0.2 eV, in LDA calculations. While the bottom of the conduction bands around the gap is relatively flat, the top of the highest valence band shows a sharp dispersive character at Γ. Figure 3 illustrates the computed band structure of $BaCo_4Sb_{12}$, and its dependence on volume. The bottom of the conduction manifold is dominated by a filler-derived s-band everywhere except near Γ, where it retains the character of anti-bonding between Co-d and Sb-s. The top of the valence band at Γ derives from an anti-bonding of Sb p-states around Sb rings [7,8]. Our band gap, E_g, is 0.225 eV for unfilled $CoSb_3$, which is comparable to the value obtained with previous calculations [8-11]. The band gaps of other configurations and fillers are summarized in Figure 3. We note that band gaps computed with LDA tend to be smaller than experimental band gaps [12].

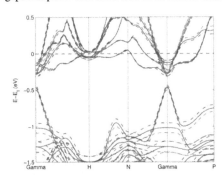

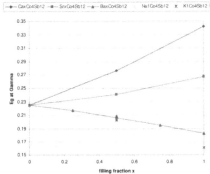

Figure 3. Band structure of $BaCo_4Sb_{12}$. Solid line: $a_0 = 9.09$Å (equilibrium), dashed: $a_0 = 9.00$ Å (P = 31.7 kbar), dash-dot: $a_0 = 9.18$ Å (P = -27.1 kbar).

Figure 4. LDA band gap (eV) at Γ as a function of filling fraction and filler element. Na and K-filled compounds are only showed for fully filled cases.

117

At the zone center, both the top of the highest valence band and the bottom of the conduction band are derived from anti-bonding states. These energy levels are known to be sensitive to the distances between participating atoms, and can thus be affected by the local distortion introduced by filling. This in turn may result in a change of the band gap at Γ. This hypothesis is tested by computing band structures under pressure. In Figure 4, a clear trend is recognized: structures under high pressure exhibits a smaller lattice size than that at equilibrium, which changes the amount of covalent overlap, resulting in a greater band gap at Γ. The trend is shared by the unfilled structure and the Ba-filled one (shown), hinting that this is independent from a specific chemistry of a filler. Further structural analysis indicates that global pressure as well as any un-relaxed local compressive stress due to a filler acts to increase the band gap. In addition, introducing extra electrons into a system increases the band gap. There is also a more direct effect of the filling which is created by charge transfer from the electropositive filler to the $CoSb_3$ structure. As seen on Figure 4, the band gaps in alkaline metal filled compounds are predicted to be lower than those containing alkaline-earth fillers of similar ionic size.

We can estimate transport properties of the material starting from the band structure by using a semiclassical Boltzmann transport approach under the constant relaxation time approximation. Figure 5 shows the computed Seebeck coefficient for a lightly-doped n-type $CoSb_3$ by using the band structure of unfilled $CoSb_3$. The carrier density is fixed as 4.54×10^{18} cm^{-3}, which is the corresponding value reported for the lightly-doped sample [13]. Agreement with experiment is very good, especially for low temperatures (around 300K). This is due to the fact that the Fermi-Dirac distribution at low temperature only activates the carriers in a very narrow energy window around the Fermi level, where the relaxation time can be well approximated by a constant. For comparison, we also show in Figure 5 the Seebeck coefficient of a fully Ca-filled compound ($CaCo_4Sb_{12}$) with a similar carrier density. There is noticeable improvement in the value at low temperatures, which agrees with the observation that the conduction band is flattened due to filling.

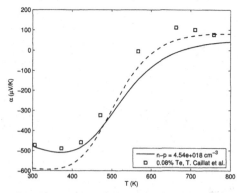

Figure 5. The Seebeck coefficient (μV/K) computed from the band structure of unfilled $CoSb_3$. Comparison is made against an experiment [13]. Dashed line: isoelectronic Ca-filled compound.

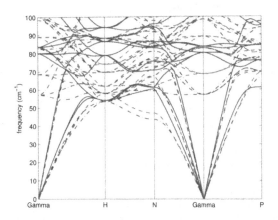

Figure 6. Vibrational spectrum unfilled $CoSb_3$ (solid), $CaCo_4Sb_{12}$ (dash-dot), and $BaCo_4Sb_{12}$ (dashed).

In order to understand the effect of the filler on the thermal transport in skutterudites, we start by calculating the vibrational spectrum using DFPT, which allows us to analyze the full phonon dispersion information in the Brillouin zone. In Figure 6 we present the effects of Ca and Ba filling on the low-frequency phonon modes. We identify the filler-dominated modes as relatively flat manifolds, around 70 cm^{-1} for Ba and 60 cm^{-1} for Ca. The weak dispersion of these bands points to the localized nature of these vibrations, i.e. that the fillers are rather decoupled. This could be advantageous when one considers the fillers' role as random scattering centers for acoustic phonons that carry thermal current. There is also clear indication that the acoustic mode frequencies are reduced by addition of filler atoms, as observed in earlier calculations. This also works to slow the heat propagation in the crystal. The bare frequency of the filler zone-center vibrations was calculated earlier to be about 90 cm^{-1} [14], assuming only the filler motion in its harmonic potential. Our full phonon calculation shows a lower frequency, which indicates significant coupling to the Sb-dominated modes. It is also interesting to note that the Ca-dominated lowest optical mode frequency is lower than that of the Ba mode, despite the atomic mass difference. This result emphasizes the importance of the specific filler atom's coupling to the Sb cage, which is dictated by the atomic size and amount of charge transfer.

CONCLUSIONS

Filling introduces only local changes in the atomic structure. Soft Sb rings accommodate the distortion and prevent its long-range propagation. Electronic band structure, including the gap at Γ, is sensitive to local atomic structure distortions, as well as volume. The direct band gap decreases when the volume increases. In our phonon calculations we observe non-trivial aspects of the fillers' vibrational dynamics. The filler-dominated vibrational modes are mostly localized (decoupled from the other fillers) and coupled with nearby Sb atoms. Also, specific chemical bonding of a filler atom with the cage modifies the conventional interpretation of the filler as an inert heavy rattler.

REFERENCES[1]

1. S. Baroni, S. de Gironcoli, A. Dal Corso, and P. Giannozzi. *Reviews of Modern Physics* **73**, 515–562 (2001).
2. S. Baroni, A. Dal Corso, S. de Gironcoli, P. Giannozzi, C. Cavazzoni, G. Ballabio, S. Scandolo, G. Chiarotti, P. Focher, A. Paswuarello, K. Laasonen, A. Trave, R. Car, N. Marzari, and A. Kokalj, Quantum-ESPRESSO. http://www.pwscf.org/
3. G.K.H. Madsen and D.J. Singh, *Computer Physics Communications* **175**, 67–71 (2006).
4. H. Takizawa, K. Miura, M. Ito, T. Suzuki, and T. Endo. *Journal of Alloys and Compounds* **282**, 79–83 (1999).
5. R. Wentzcovitch, VLAB. http://www.vlab.msi.umn.edu/
6. T. Schmidt, G. Kliche, and H.D. Lutz, *Acta Crystallographica (Section C)-Crystal Structure Communications* **43**, 1678–1679 (1987).
7. D.W. Jung, M.H. Whangbo, and S. Alvarez, *Inorganic Chemistry* **29**, 2252–2255 (1990).
8. J.O. Sofo and G.D. Mahan, *Physical Review B* **58**, 15620–15623 (1998).
9. Philippe Ghosez and Marek Veithen, *Journal of Physics-Condensed Matter* **19**, 096002 (2007).
10. E.Z. Kurmaev, A. Moewes, I.R. Shein, L.D. Finkelstein, A.L. Ivanovskii, and H. Anno, *Journal of Physics-Condensed Matter* **16**, 979–987 (2004).
11. I. Lefebvre-Devos, M. Lassalle, X. Wallart, J. Olivier-Fourcade, L.R. Monconduit, and J.C. Jumas, *Physical Review B* **63**, 125110 (2001).
12. R. M. Martin, *Electronic Structure: Basic Theory and Practical Methods*, Cambridge University Press (2004).
13. T. Caillat, A. Borshchevsky, and J.P. Fleurial, *Journal of Applied Physics* **80**, 4442–4449 (1996).
14. J. Yang, W. Zhang, S.Q. Bai, Z. Mei, and L. Chen, *Applied Physics Letters* **90**, 192111 (2007).

---

[1] At the date this paper was written, URLs or links referenced herein were deemed to be useful supplementary material to this paper. Neither the author nor the Materials Research Society warrants or assumes liability for the content or availability of URLs referenced in this paper.

Mater. Res. Soc. Symp. Proc. Vol. 1166 © 2009 Materials Research Society   1166-N05-08

Transport Properties of Thermoelectric Nanocomposites

L.M. Woods, A. Popescu, J. Martin, and G.S. Nolas
Department of Physics, University of South Florida,
Tampa, FL 33620, U.S.A.

ABSTRACT

We present a theoretical model for carrier conductivity and Seebeck coefficient of thermoelectric materials composed of nanogranular regions. The model is used to successfully describe experimental data for chalcogenide PbTe nanocomposites. We also present similar calculations for skutterudite CoSb$_3$ nanocomposites. The carrier scattering mechanism is considered explicitly and it is determined that it is a key factor in the thermoelectric transport process. The grain interfaces are described as potential barriers. We investigate theoretically the role of the barrier heights, widths, and distances between the barriers to obtain an optimum regime for the composites thermoelectric characteristics.

INTRODUCTION

Thermoelectric (TE) energy conversion materials offer the possibility of an all-solid-state technology to convert heat into electrical energy. The performance of devices based on such materials is characterized by the dimensionless figure of merit defined as $ZT = S^2 \sigma T / \kappa$, where S is the Seebeck coefficient (thermopower), σ - the carrier conductivity, κ - the thermal conductivity, and T – the absolute temperature. Larger values of ZT warrant more efficient devices, thus researchers have devoted much effort in finding ways to increase the figure of merit. However, the Seebeck coefficient, carrier and thermal conductivities are interrelated in the usual bulk materials. It is not possible to influence one of them without influencing the others in some disadvantageous ways. This imposes limitations on how much the figure of merit of bulk materials can be improved. For the past several years ZT of materials currently used in commercial devices has been ~ 1 for all applicable temperature ranges[1].

Other efforts have focused on the possibility to alter the thermoelectric properties independently by approaching nanoscaled dimensions. These include Bi$_2$Te$_3$/Sb$_2$Te$_3$ supperlattices[2], quantum dot supperlattices[3], or one dimensional quantum wires[4] for which the thermal conductivity is reduced thus leading to an increased figure of merit. In addition, bulk materials with nanostructured inclusions have also been demonstrated to have enhanced thermoelectric properties. For example, studies of PbTe with Pb nanoprecipitates[5], nanocrystalline CoSb$_3$[6], and nanogranular PbTe composites[7,8] have reported an increased Seebeck coefficient as compared to the bulk counterpart. These investigations[5-8] show that the carrier scattering by interfaces present in the bulk matrix is a key factor in the improved TE properties. In fact, it has been shown[8,9] that the interface grain barriers filter the low energy carriers, while the higher energy ones diffuse through the specimen. Since the mean energy per carrier is increased, for certain conditions the Seebeck coeffient is increased while the carrier conductivity is not degraded appreciably.

Here we describe a phenomenological model describing the diffusion transport of carriers through bulk material which contains nanogranular regions. In addition to carrier-acoustic phonon, carrier-optical phonon, and carrier-impurity scattering mechanisms, we consider the carrier-interface scattering. The material is viewed as containing potential interface barriers due to the grains which do not alter the energy band structure of the bulk significantly. The model is tested by comparing the calculated and measured Seebeck coefficient and electrical conductivity for PbTe nanocomposites. We use this theory to understand the role of the grain characteristics, such as barrier height, grain size, and distance between the grains in order to determine an optimum regime of TE properties of such granular nanocomposite materials.

THEORY

We consider quasi-equilibrium and diffusive transport of charge carriers in bulk. The charge transport properties can be described by

$$\sigma = \frac{2e^2}{3m^*} \int_0^\infty \tau(E)g(E)E\left(-\frac{\partial f(E)}{\partial E}\right)dE, \quad (1)$$

$$S = \frac{1}{eT}\left[\frac{\int_0^\infty \tau(E)g(E)E^2\left(-\frac{\partial f(E)}{\partial E}\right)dE}{\int_0^\infty \tau(E)g(E)E\left(-\frac{\partial f(E)}{\partial E}\right)dE} - \mu\right] \quad (2)$$

where e is the electron charge, m^* - the effective mass, μ - the chemical potential for the specific material, $\tau(E)$ - the momentum relaxation time for the charge carriers, $g(E)$ - the total density of states (DOS) for the material, and $f(E)$ - the energy distribution function. We consider the energy dependent functions explicitly $g(E)$, $f(E)$, and $\tau(E)$ and explain the relevant assumptions made in the model.

Total Density of States We assume that the grains serve only as scattering barriers to the charge carriers and do not affect the energy band structure inside the grains. Thus the total density of states (DOS) is taken to be the same as the bulk material. For several small band-gap TE semiconductors, such as lead chalcogenides and skutterudites, the non-parabolic two-band Kane model is a good approximation for the energy bands involved in the transport[10, 11]. The total DOS for the two-band Kane model can be written as[10]

$$g(E) = \frac{\sqrt{2}}{\pi^2}\left(\frac{m^*}{\hbar^2}\right)^{3/2}\sqrt{E(1+E/E_g)}(1+2E/E_g) \quad (3)$$

where $\hbar$ is the Plank's constant, E_g - the energy band gap, and m^* - the effective mass for all directions.

Energy Distribution Function For diffusive and quasiequilibrium transport the energy distribution function is taken to be the Fermi function $f(E) = 1/(e^{(E-E_F)/k_B T} + 1)$ with E_F being the Fermi level, and k_B - the Boltzmann constant. The Fermi level is specific for each material and it is also related to the charge carrier concentration

$$p = \frac{4}{\sqrt{\pi}}\left(\frac{m^* k_B T}{2\pi\hbar^2}\right)^{3/2}\int_0^\infty E^{1/2}f(E)dE \quad (4)$$

The self-consistent solution of p allows one to determine the Fermi level for a specific carrier concentration.

122

Scattering Mechanisms The total relaxation time is obtained from the Mathiessen's rule $\frac{1}{\tau(E)} = \sum_i \frac{1}{\tau_i(E)}$, where $\tau_i(E)$ is the relaxation time for a specific scattering mechanism. In this model we take into account carrier scattering from the grain interface barrier $\tau_b(E)$, acoustic phonons $\tau_{a-ph}(E)$, optical phonons $\tau_{o-ph}(E)$, and ionized

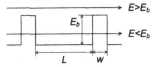

Fig. 1 Potential barriers with height E_b and width w separated at a distance L from each other. Carriers with energy E impede on the barriers.

impurities $\tau_{imp}(E)$. The carrier-grain interface scattering is described by assuming that the material consists of an infinite number of grain boundaries modeled as potential barriers of equal height E_b, width w, and space between two adjacent barriers L – Fig. 1. The average values of E_b, w, and L can be taken from experimental measurements[7,12]. The diffusion process is described by considering the quantum mechanical transmission probability $T(E)$ of carriers with energy E impeding on the barriers. The carrier path length λ after passing through the first barrier is $T(E)(1-T(E))L$, after passing through the second barrier is $T(E)^2(1-T(E))2L$, and finally after passing through N barriers is $T(E)^n(1-T(E))nL$. Thus assuming that the mean free path due to this scattering mechanism is simply λ, for infinite number of barriers one finds that

$$\lambda = \sum_{n=1}^{N \to \infty} T(E)^n (1-T(E))nL = \frac{T(E)L}{1-T(E)}, \quad \tau_b(E) = \frac{\lambda}{v} \qquad (5)$$

where $v = \sqrt{\frac{2E}{m^*}}$ is the average velocity of the charge carriers. The transmission probability $T(E)$ can also be obtained using standard quantum mechanics by taking the incident carriers upon the barriers as plane wave-like[9].

The scattering mechanisms from acoustic and optical phonons and ionized impurities can be described using available expressions. The relevant relaxation times therefore are taken to be

$$\tau(E)_{a-ph} = \frac{h^4}{8\pi^3} \frac{\rho v_L^2}{k_B T} \frac{1}{(2m^*)^{3/2} D^2} E^{-1/2},$$

$$\tau(E)_{o-ph} = \frac{h^2}{2^{1/2} m^{*1/2} e^2 k_B T (\varepsilon_\infty^{-1} - \varepsilon_0^{-1})} E^{1/2} \qquad (6)$$

$$a_{imp} = \left[\frac{Z^2 e^4 N_i}{16\pi(2m^*)^{1/2} \varepsilon^2} \ln\left[1 + \left(\frac{2E}{E_m}\right)^2 \right] \right]^{-1} E^{3/2}$$

where ρ is the mass density, v_L - the longitudinal velocity of sound, D - the deformation potential constant, ε_∞ - the high frequency dielectric constant, ε_0 - the static dielectric constant, and ε - the dielectric constant of the medium. Also, $E_m = \frac{Ze^2}{4\pi\varepsilon r_m}$ is the potential energy at a distance r_m from an ionized impurity with r_m approximately half the mean distance between two adjacent impurities, and N_i is their concentration. These quantities can be found from available data in the literature or estimated experimentally for a particular specimen.

123

DISCUSSION

The model described here provides all the necessary tools to calculate and explain the experimental data for σ and S for small band gap semiconducting TE materials containing granular interfaces. Its applicability lies within the assumptions that the carrier transport is diffusive, the relaxation time approximation is valid, the electronic band structure within the grains is the same as in bulk, and that it can be described using the two-band Kane model. Here we test this theory to see how effectively it can describe experimental data for PbTe granular nanocomposites. All approximations regarding the carrier relaxation time and electronic structure assumptions are valid for this material.

Ag-doped PbTe nanocrystals were synthesized as described previously[13]. The synthesis approached allowed for reproducible 100 - 150 nm spherical PbTe nanocrystals, also confirmed by TEM, with a high yield of over 2 grams per batch. The nanocrystals were subjected to Spark Plasma Sintering to achieve ~ 95 % bulk theoretical density, resulting in a dimensional nanocomposite structure. Low temperature transport measurements were performed on the specimens. The thermoelectric properties of these materials are reported in another article in this volume[14].

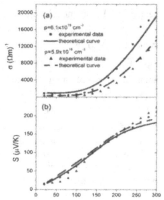

Fig. 2 Calculated and measured (a) electrical conductivity and (b) Seebeck coefficient as a function of temperature T for two values of the carrier concentration.

The experimental results for the carrier conductivity and Seebeck coefficient for two specimens are given in Fig. 2. The experimental results are compared with the calculations for S and σ, which are also shown in Fig. 2. The agreement is excellent. The calculations were done without the carrier-ionized impurity scattering since the presence of such impurities was found insignificant in the experiments. We used the following barrier parameters estimated from the data - E_b=60 meV, w=50 nm, and L=300 nm. The other physical parameters for PbTe are also listed in Table 1. The model we have developed and tested for relevant measurements for PbTe materials can further be used to elucidate the role of factors such as carrier concentration, electronic structure parameters, and grain interface characteristics. For example, one can calculate the conductivity, Seebeck coefficient, and power factor $S^2\sigma$ using Eqs. (1, 2) as a function of E_b, w, and/or L for a specific material and carrier concentration p. In fact, such calculations are useful to understand and quantify the carrier filtering process due to the carriers scattering from the granular interfaces.

We find that by keeping all parameters the same and changing E_b, leads to particular changes in the TE transport properties. Increasing E_b leads to a decrease in the conductivity and an increase in the Seebeck coefficient. This is explained by realizing that higher barriers will

Table 1 Properties of chalcogenide PbTe and skutterudite CoSb₃ materials.

Material	$\rho \ (g/cm^3)$	$v_L \ (m/s)$	$D \ (eV)$	m^*/m_0	$\varepsilon_\infty^{-1} - \varepsilon_0^{-1}$
PbTe	8.16	1730	5 – 15	0.16	0.0022
CoSb₃	7.64	4590	4	0.07	0.011

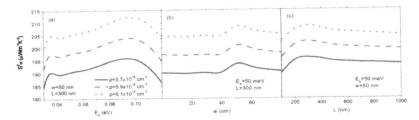

Fig. 3 Power factor for PbTe as a function of (a) barrier height E_b; (b) barrier width w; and (c) distance between the barriers L for various carrier concentrations and T=300 K.

scatter more electrons, therefore the number of carriers contributing to the charge transport is decreased. Since fewer carriers contribute to the transport, the σ is decreased. The same process will contribute to an increased mean energy per carrier, thus the Seebeck coefficient is increased. Similar behavior can be achieved by changing w or L. Increasing w results in a smaller transmission $T(E)$ through the barriers. Thus σ is decreased and S is increased again. If L is decreased, there will be more frequent carrier scattering events, which also results in a decreased σ and increased S.

Experimentally, one may be able to change different parameters characterizing the potential grain barriers in order to manipulate the carrier transport characteristics as a function of temperature. For thermoelectric applications, however, one is interested in increasing the power factor $S^2\sigma$. Thus optimal E_b, w, and/or L are needed to achieve such σ and S in order to obtain the highest performing TE materials. Using Eqs. (1, 2) we can calculate $S^2\sigma$ as a function of E_b, w, and L at a specific temperature. Fig. 3 shows our results for the power factor for PbTe composites as a function of the grain characteristics. All scattering mechanisms are included and the impurity concentration is taken to be $N_i = 15\%\,p$. The fact that $S^2\sigma$ exhibits maxima shows that the influence of the grains may or may not be beneficial for the TE transport. If more charge carriers (lower E_b), less frequent scattering (longer L), and larger transmission through the barrier (smaller w) are allowed, the electrical conductivity increases. In contrast, when the mean energy per carrier is increased by taking higher barriers (larger E_b), more frequent scattering (smaller L), and smaller transmission (larger w), the Seebeck coefficient increases. The key is to use grains with such properties that balance the interplay between increasing S and decreasing σ in order to obtain an optimum power factor.

We also present results from calculating the transport properties at different temperatures of the skutterudite $CoSb_3$ – Fig. 4. The structural parameters are in Table 1. All scattering mechanisms are included and $N_i = 15\%\,p$. Fig. 4a) and b) show that $\sigma^2 S$ has similar behavior as the one for PbTe. In particular, the power factor exhibits maxima for certain values of E_b and w. When L is changed, however, $\sigma^2 S$ increases towards its value for bulk.

CONCLUSIONS

In conclusion, we have presented a model to calculate carrier transport properties for thermoelectric materials containing nanogranular regions. The model was tested successfully for appropriate experimental measurements of PbTe composites. Our calculations for PbTe and

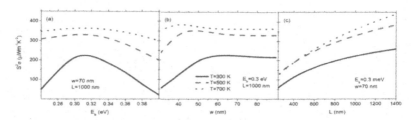

Fig. 4 Power factor for CoSb$_3$ as a function of (a) barrier height E_b; (b) barrier width w; and (c) distance between the barriers L for different temperatures and $D=6.2 \times 10^{18}$ cm^{-3}.

CoSb$_3$ show that the interplay between the grain interface scattering together with other mechanisms due to carrier–phonon and carrier–impurity scattering mechanisms can give a set of granular parameters which result in optimum thermoelectric properties.

ACKNOWLEDGMENTS

The authors acknowledge support by the US Army Medical Research and Material Command under Award No. W81XWH-07-1-0708. Opinions, interpretations, conclusions, and recommendations are those of the authors and are not necessarily endorsed by the U.S. Army.

REFERENCES

1. M.S. Dresselhaus *et al*, "New directions for low dimensional thermoelectric materials", Adv. Mater. **19**, 1042 (2007); G.J. Snyder and E.S. Toberer, "Complex Thermoelectric Materials", Nature Mater. **7**, 105 (2008);
2. R. Venkatasubramanian, E. Siivola, T. Colpitts, and B. O'Quinn, Nature **413**, 597 (2001);
3. T.C. Harman, P.J. Taylor, M.P. Walsh, and B.E. LaForge, Science **297**, 2229 (2002);
4. L.D. Hicks and M.S. Dresselhaus, Phys. Rev. B **47**, 16631 (1993);
5. K.F. Hsu, S. Loo, F. Guo, J.S. Dyck, C. Uher, T. Hogan, E.K. Polychroniadis, and M.G. Kanatzidis, Science **303**, 818 (2004);
6. M.S. Toprak, S. Stiewe, D. Platzek, S. Williams, L. Bertini, E. Mueller, C. Gatti, M. Rowe, amd M. Muhammed, Adv. Func. Mat. **14**, 1189 (2004);
7. J. Martin, G. S. Nolas, W. Zhang, and L. Chen, Appl. Phys. Lett. **90**, 222112 (2007);
8. J.P. Heremans, C.M. Thrush, and D.T. Morelli, Phys. Rev. B **70**, 115334 (2004);
9. A. Popescu, L.M. Woods, J. Martin, and G.S. Nolas, Phys. Rev. B, to be published;
10. D. M. Rowe and C. M. Bhandari, *Modern Thermoelectrics* (London, Holt Saunders, 1983);
11. J.O. Sofo and G.D. Mahan, Phys. Rev. B **58**, 15620 (1998);
12. K. Kishimoto and T. Koyanagi, J. Appl. Phys. **92**, 2544 (2002); K. Kishimoto, K. Yamamoto, and T. Koyanagi, Jpn. J. Appl. Phys. **42**, 501 (2003).
13. J. Martin, L. Wang, L. Chen, and G. S. Nolas, Phys. Rev B **79**, 115311 (2009);
14. H. Kirby, J. Martin, A. Datta L. Chen, and G. S. Nolas, Mater. Res. Soc. Symp., present volume.

Mater. Res. Soc. Symp. Proc. Vol. 1166 © 2009 Materials Research Society   1166-N05-09

Enhancement of the Thermoelectric Figure of Merit in Gated Bismuth Telluride Nanowires

Igor Bejenari[1], Valeriu Kantser[2] and Alexander A. Balandin[1]
[1]Department of Electrical Engineering and Materials Science and Engineering Program, University of California – Riverside, Riverside, California 92521 U.S.A.
[2]Institute of Electronic Engineering and Industrial Technologies, Academy of Sciences of Moldova, Kisinev, MD 2028 Moldova

ABSTRACT

We theoretically studied how the electric filed effect can modify thermoelectric properties of intrinsic bismuth telluride nanowires, which are grown along [110] direction. The electronic structure and wave functions were calculated by solving the self-consistent system of the Schrodinger and Poisson equations by means of both the Thomas-Fermi approximation and the spectral element method. The thermoelectric parameters were determined using a constant relaxation-time approximation. The external electric field can increase the Seebeck coefficient of a nanowire with 7 - 15 nm lateral dimensions by nearly a factor of two, and enhance the figure of merit by an order of magnitude.

INTRODUCTION

Bismuth telluride and its solid solutions ($Bi_{2-x}Sb_xTe_3$, $Bi_2Te_{3-y}Se$) are one of the best known thermoelectric materials for the modern commercial application. These materials possess notable properties like a high anisotropic multi-valley Fermi surface, a small value of the thermal conductivity, and an optimal value of a carrier concentration at room temperature. Because the electron dimensional confinement and the enhancement of the photon boundary scattering control the electron transport, bismuth telluride based superlattices and quantum wires have advanced thermoelectric properties when compared to bulk bismuth telluride. For example, the figure of merit of the Bi_2Te_3/Sb_2Te_3 superlattices achieves its maximum value of 2.4 at room temperature [1]. Because of a larger confinement effect, for thermoelectric applications the bismuth telluride nanowires are better than the superlattices. The bismuth telluride nanowires can be obtained by electrochemical deposition of the material in the nanopores of anodized alumina membranes, by Taylor-Ulitovsky technique, and by high pressure injection of the melt into capillaries [2, 3].

Electric field effect (EFE) is a powerful tool to control electrical properties of low dimensional structures. Experimental and theoretical studies showed that the EFE can significantly improve thermoelectric properties of the Bi nanowires and PbTe films [4-6]. While dimensional confinement of electrons leads to a modification of their density of states, the Fermi level can be changed due to the EFE control of the electron (hole) concentration in the nanowire. The EFE on the nanowire thermoelectric properties is also a result of the energy spectrum modification and the local dependence of the electron distribution function. Also, the applied electric field causes additional quantum confinement. The EFE on the nanowire thermoelectric properties is also due to the energy spectrum modification because of the additional quantum confinement caused by the applied electric field and the local dependence of the electron distribution function. These factors, affected by the side gate potential, lead to an improvement of the nanowire thermoelectric properties.

In this paper we theoretically studied how the electric field effect influences the thermoelectric properties of the intrinsic bismuth telluride nanowires that have square cross sections of 7x7 and 15x15 nm^2 at room temperature. The electron mobility in the nanowires is supposed to coincide with the electron mobility in the bismuth telluride bulk material. We used the carrier effective mass components calculated previously for the nanowire with growth direction of [110] [7]. The transport parameters were estimated using the semi-classical Boltzman approximation [8]. We used a constant-relaxation-time approximation to calculate the Seebeck coefficient, and electrical and thermal conductivities. The justification of applying this approach to the calculation of the transport properties of bismuth telluride nanowire material is given in our previous paper [7].

THEORY

We considered the Bi_2Te_3 nanowire connected to the source and the drain and separated from the lateral metallic gate by a dielectric SiO_2 layer with a thickness d_{SiO2}=300 nm. A similar device based on the Bi nanowire with a thickness of 28 nm has been experimentally studied recently [6]. The following derivations were applied. To estimate the nanowire transport properties we found a solution of a self-consistent system of Schrodinger and Poisson equations

$$-\frac{\hbar^2}{2m_x}\frac{d^2\Psi}{dx^2}-\frac{\hbar^2}{2m_z}\frac{d^2\Psi}{dz^2}+[V_c(x,z)+E_c-e\varphi(x,z)]\Psi(x,z)=E\Psi(x,z),\qquad(1)$$

$$-\varepsilon\varepsilon_0\Delta\varphi(x,z)=\rho(x,z).\qquad(2)$$

where E_c is the bottom of the conduction band, φ is the electrostatic potential, and V_c is the confinement potential which corresponds to an infinitely high cylindrical quantum well. After replacing the sign in front of the second derivatives in equation 1 by an opposite one, we obtained the Schrodinger equation for holes. The applied electric field was assumed to be directed along the z axis. We did not take into considerations the local exchange-correlation energy because the electron (hole) density in the intrinsic semiconductor nanowires is supposed to be small. Integrating the both sides of the Poisson equation, we obtained the charge neutrality equation for the nanowire in the form [9]

$$n_{1D}(E_F)=p_{1D}(E_F)+a_x\sigma.\qquad(3)$$

Here the surface electric charge $-e\sigma$, induced by the gate electric field in the interface SiO_2/Bi_2Te_3, is associated with a Fermi-level-pinning boundary condition. Condition (3) means that the number of electrons and holes populating the subbands is equal to the number of charge carriers occupying surface states. The surface states do not contribute to the transport because of the strong surface roughness scattering. The applied electric field induces a surface charge on the SiO2/Bi2Te3 interface. The effective (induced) surface charge doping concentration σ can be estimated in the gate–SiO_2–Bi_2Te_3 capacitance structure as a function of the gate voltage V_g [6]

$$\sigma = \frac{\varepsilon_0 \varepsilon_{SiO2}}{ed_{SiO2}} V_g .$$ (4)

Where ε_{SiO2}=3.9 is the permittivity of the SiO$_2$ layer. To solve the system of equations 1 and 2 we use both the Thomas-Fermi (TF) approximation and the Spectral Element Method (SEM) [9-11]. In the Thomas-Fermi approximation, the mathematical definitions of the Seebeck coefficient, the electrical and thermal coefficients, and the figure of merit coincide with those given for the nanowires in the absence of the gate voltage [7]. In the framework of the SEM, the local dependence of the transport parameters is defined by the electron (hole) wave function [9]. For example, using the notations from our previous paper [6], the electron concentration can be written as

$$n_{1D}(x, z; E_F) = N_{c,v}^{1D} \sum_{n=1}^{n_{max}} \sum_{l=1}^{l_{max}^{(n)}} \Phi_{-\frac{1}{2}}(\eta_{n,l}) |\Psi(x, z)|^2 .$$ (5)

Here, factor $N_{c,v}^{1D} = \left(2m_y^{e,h} k_B T / \pi \hbar^2\right)^{\frac{1}{2}}$ denotes the effective density of electron (hole) states. The thermoelectric parameters are effectively macroscopic parameters. Hence, with a side gate electric field, the figure of merit ZT and the Seebeck coefficient S are defined by averaging as [5]

$$ZT = \frac{\langle \sigma(z) S(z) \rangle^2}{\langle \sigma(z) \rangle \langle \kappa(z) \rangle} T ,$$ (6)

$$S = \frac{\langle \sigma(z) S(z) \rangle}{\langle \sigma(z) \rangle} .$$ (7)

DISCUSSION

Figure 1 shows the dependence of the electron and hole concentrations in the nanowire on the gate voltage at room temperature. The results obtained using the TF approximation and obtained using the SEM agree well. At the zero gate voltage, the charge carrier concentrations in the bismuth telluride nanowires are $n = p = 3.11 \times 10^{-16}$ cm^{-3} and $n = p = 3.03 \times 10^{-17}$ cm^{-3} for the thicknesses of 7 and 15 nm, respectively. Under an applied electric field, the carrier concentration increases by several orders of magnitude. At the positive (negative) gate voltage the electrons (holes) dominate. Hence, it is possible to manage the type of nanowire conductivity by applying the gate voltage. This fact was experimentally confirmed for the Bi nanowires [6].

129

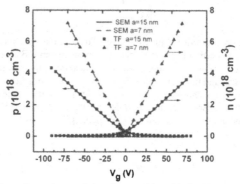

Figure 1. The gate voltage dependence of the electron (n) and hole (p) concentrations of the Bi_2Te_3 nanowires with the thicknesses a=7 nm (dashed line) and 15 nm (solid line) at room temperature.

The gate voltage dependence of the carrier concentration increases with a decrease of the nanowire thickness. For the nanowire thickness of 15 nm, the dependence of the electron concentration on the gate voltage is less pronounced compared to that for holes. This is, because the splitting between the energy subbands decreases for nanowires with large thicknesses due to a lesser confinement effect. Thus the number of subbands, involved in the transport, decreases with a decrease of the nanowire thickness. Since the hole effective mass is greater than the electron effective mass, the hole concentration increases faster with the gate voltage compared to the electron concentration.

Figure 2 shows the dependence of the nanowire Seebeck coefficient on the gate voltage. For the nanowire with the thickness of 15 nm, a small disagreement between the TF and SEM data appears at the large voltage, owing to the great number of the subbands involved in the transport. At the zero gate voltage, the Seebeck coefficient is equal to -167 and -252 μV/K for the nanowire thicknesses of 15 and 7 nm correspondingly. For the nanowire thickness of 7 nm, the maximum absolute value of the Seebeck coefficient approaches the value of 478 μV/K (494 μV/K) at the gate voltage equal to -3 V (2 V). For the thickness of 15 nm, the maximum absolute value of 333 μV/K (344 μV/K) is achieved at the greater gate voltage equal to -29 V (15 V). Since the electron mobility in the bismuth telluride is equal to 1200 cm^2/Vs and it is more than twice larger than the hole mobility, which is equal to 510 cm^2/Vs [7], the maximum absolute values of the Seebeck coefficient corresponding to the opposite gate polarities are achieved at the different gate voltages. The EFE causes the Seebeck coefficient to increase as much as twice. Thus both the confinement effect and the electric field effect considerably improve the nanowire Seebeck coefficient. The calculations showed that the dependence of the nanowire thermal conductivity on the gate voltage is not monotonic. At the zero gate voltage, the thermal conductivity is equal to 0.248 $WK^{-1}m^{-1}$ (0.702 $WK^{-1}m^{-1}$) for the nanowire thickness 7 nm (15 nm). The minimum value of the thermal conductivity of 0.18 $WK^{-1}m^{-1}$ (0.39 $WK^{-1}m^{-1}$) is achieved at the gate voltage of 2 and -5 V (30 and -50 V) for the nanowire thickness of 7 nm (15 nm).

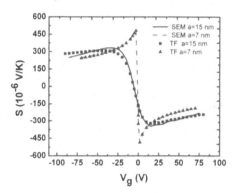

Figure 2. The gate voltage dependence of the Seebeck coefficient S of the Bi_2Te_3 nanowires with the thicknesses a=7 nm (dashed line) and a=15 nm (solid line) at room temperature.

Figure 3 presents the gate dependence of the nanowire figure of merit at room temperature. At the zero gate voltage, the figure of merit is equal to 0.065 and 0.099 for the nanowire thicknesses of 7 and 15 nm, respectively. For the nanowire thickness of 7 nm (15 nm), the maximum value of the figure of merit ZT=3.4 (2.3) is achieved at the gate voltage 15 V (40 V). For the opposite gate polarity, when the holes dominate, the maximum value of the figure of merit achieves the lower value, because of the difference between the hole mobility and the electron mobility, as well as due to the difference in the effective masses of electrons and holes. Since the mathematical expression for the nanowire figure of merit is rather complex, the discrepancy between the TF and the SEM data increases with the gate voltage for the nanowire thickness of 15 nm.

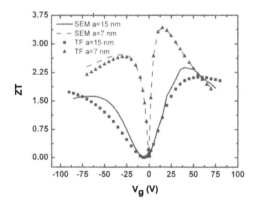

Figure 3. The gate voltage dependence of the figure of merit ZT of the Bi_2Te_3 nanowires with the thickness a=7 nm (dashed line) and a=15 nm (solid line) at room temperature.

CONCLUSIONS

The comparison of the results obtained by the SEM and TF approximations shows that the TF approximation can be successfully used for the calculation of the electric field effect on the transport properties of nanowires with small thicknesses. The EFE can effectively control the type of the nanowire conductivity. Both the confinement effect and the electric field effect considerably improve the intrinsic nanowire thermoelectric properties. The EFE increases the Seebeck coefficient nearly as much as twice. The maximum value of the figure of merit achieves values of 3.4 and 2.3 for the nanowire thicknesses of 7 and 15 nm, correspondingly.

ACKNOWLEDGMENTS

I.B. acknowledges financial support of the Fulbright Scholar Program. A.A.B. acknowledges support from the SRC – DARPA Functional Engineered Nano Architectonics (FENA) center. We wish to thank D. Kotchetkov for his careful reading of our paper and for the helpful comments.

REFERENCES

1. R. Venkatasubramanian, E. Siivola, T. Colpitts, and B. O'Quinn, *Nature* **413**, 597 (2001)
2. J. Zhou, C. Jin, J. H. Seol, X. Li, and L. Shi, *Appl. Phys. Lett.* **87**, 133109 (2005).
3. L. Li, Y. Yang, X. Huang, G. Li, and L. Zhang, *Nanotechnology* **17**, 1706 (2006).
4. A. V. Butenko, V. Sandomirsky, Y. Schlesinger, and Dm. Shvarts, *J. Appl. Phys.* **82**, 1266-1273 (1997).
5. V. Sandomirsky, A. V. Butenko, R. Levin, and Y. Schlesinger, *J. Appl. Phys.* **90**, 2370 (2001).
6. Boukai, K. Xu, and J.R. Heath, *Adv. Mater.* **18**, 864 (2006).
7. I. Bejenari, V. Kantser, *Phys. Rev. B* **78**, 115322 (2008).
8. V. I. Zubkov, *Fiz. Tekh. Poluprovodn.* **40**, 1236 (2006) [Sov. Phys. *Semicond.* **40**, 1204 (2006)].
9. J. H. Luscombe, A. M. Bouchard, and M. Luban, *Phys. Rev. B* **46**, 10262 (1992).
10. B. K. Ridley, *Quantum Processes in Semiconductors*, 3rd ed. (Oxford Univ. Press Inc., New York, 1993) p. 378.
11. C. Cheng, Q. H. Liu, J.-H. Lee, and H. Z. Massoud, *J. Comput. Electron.* **3**, 417 (2004).

Bulk Materials

Mater. Res. Soc. Symp. Proc. Vol. 1166 © 2009 Materials Research Society   1166-N03-16

Syntheses of Ni-Doped and Fe-Doped CoSb$_3$ Thermoelectric Nanoparticles Through Modified Polyol Process

Takashi Itoh[1] and Keisuke Isogai[1]
[1]EcoTopia Science Institute, Nagoya University, Furo-cho, Chikusa-ku, Nagoya, 464-8603, Japan

ABSTRACT

Skutterudite CoSb$_3$ compounds are of increasing interest as materials with good thermoelectric performance over the temperature range of 600 to 800 K, but the thermal conductivity of the materials is relatively high. Nanostructured materials have been shown to enhance phonon scattering and lower the thermal conductivity of the thermoelectric materials. Partial substitution of Ni or Fe on the Co site of CoSb$_3$ compounds is a hopeful route for improving thermoelectric performance of the CoSb$_3$ compounds. In the present work, synthesis of Ni-doped and Fe-doped CoSb$_3$ nanoparticles through the modified polyol process was attempted and the optimum synthesizing condition was investigated. Co(OOCH$_3$)$_2 \cdot$4H$_2$O, Ni(OOCH$_3$)$_2 \cdot$4H$_2$O, FeCl$_3 \cdot$6H$_2$O and SbCl$_3$, were prepared as precursors. The precursors were reduced by NaBH$_4$ in tetraethyleneglycol at 513 K in an argon atmosphere, for different reaction times (holding times). The reaction products were characterized by the X-ray diffraction, the energy dispersive X-ray spectroscopy, and transmission electron microscopy. The nanoparticles with about 20 to 30 nm in size mainly existed in the reaction products regardless of the chemical composition and the reaction time. The skutterudite phase was identified as a main phase in the sample synthesized for long reaction time, but the other phases of Sb and MSb$_2$ (M=Co, Ni, Fe) were also detected. The lattice parameter of the synthesized skutterudite phase linearly increased with increasing the doping agent concentration, following Vegard's law.

INTRODUCTION

More than 60 % of heat energy obtained by burning fossil fuels such as petroleum, natural gas, and coal is discharged as waste heat during energy conversion in energy systems such as automobiles, combined cycles and waste material incineration systems. Thermoelectric power generation has attracted attentions because it can directly convert the waste heat to electric energy through the thermoelectric semiconductor modules. It is expected to contribute the resolution of environmental problems and the energy security.

Skutterudite compounds such as CoSb$_3$ are well known to have the relatively good thermoelectric properties with respect to Seebeck coefficient and electrical resistivity. However, thermoelectric performance of the CoSb$_3$ compound is not improved due to the relatively high thermal conductivity. As an approach for reduction of the lattice thermal conductivity, the crystal structure has been modified into the filled-skutterudite structure, in which the rare-earth ions are incorporated as "rattlers" into the interstitial sites of lattice [1-3]. The rattling behavior of the rare-earth ions causes phonon scattering by "guest ion"-phonon coupling [4], reducing the lattice thermal conductivity. Reduction of phonon mean free path by decreasing grain size of the skutterudite compound and by uniformly dispersing nanoparticles into matrix of the compound could bring effective reduction of the lattice thermal conductivity, leading to improvement in thermoelectric properties [5].

Nanoparticles are very effective as starting raw powder for making a super-fine grain structure through a sintering process. There are a lot of production routs of nanoparticles, i. e. ball milling, chemical vapor deposition, gas-phase condensation, thermal decomposition, high-temperature/high-pressure processing, solution-based processing, etc. Synthesis of nanoparticles using the solution-based route has several advantages to precipitate nanocrystalline powders at low temperatures and to control over nanostructure and morphology.

Fievet et al. [6] originally developed a solution-based route, namely "polyol process", for synthesis of submicron-size metal particles. In the process, a polyalcohol with relatively high-boiling point is used as both a solvent and a mild reducing agent. Then, Schaak et al. [7] have modified the process by adding a powerful reducing agent to produce size- and shape-controlled nanocrystals of a variety of metals, alloys and intermetallices. This approach can avoid a high-temperature melting step that is often required in traditional bulk syntheses and greatly shorten the time required to form them. In our previous work [8], synthesis of the $CoSb_3$ nanoparticles via the modified polyol process was attempted, and the influences of the reaction time and the total amount of raw materials on the synthesized nanoparticles were investigated. Hng et al. have also reported about the synthesis of $CoSb_3$ via the modified polyol process [9]. They used a reactant (cobalt dichloride hexahydrate) different from that (cobalt acetate tetrahydrate) in our previous work. In this research, the synthesis of Ni-doped and Fe-doped $CoSb_3$ nanoparticles through the modified polyol process was attempted and the optimum synthesizing condition was investigated.

EXPERIMENT

For synthesizing Ni-doped $CoSb_3$, prescribed amount of three precursors, cobalt acetate tetrahydrate ($Co(OOCCH_3)_2·4H_2O$), antimony trichloride ($SbCl_3$) and nickel acetate tetrahydrate ($Ni(OOCH_3)_2·4H_2O$) were dissolved in 20 ml tetraethyleneglycole (TEG) as a polyalcohol solvent together with 170 mg poly vinyl pyrrolidone (PVP) as a surface stabilizer by sonicating and then magnetic stirring. For synthesizing Fe-doped $CoSb_3$, prescribed amount of $Co(OOCCH_3)_2·4H_2O$, $SbCl_3$ and ferric trichloride hexahydrate ($FeCl_3·6H_2O$) were also dissolved in 20 ml TEG with 170 mg PVP by sonicating and then magnetic stirring. Ni content and Fe

Table I. Amount of precursors prepared for $Co_{1-x}Ni_xSb_3$ and $Co_{1-y}Fe_ySb_3$ nanoparticle syntheses.

$Co_{1-x}Ni_xSb_3$				
Precursors	x=0	x=0.03	x=0.06	x=0.1
$Co(OOCCH_3)_2·4H_2O$	1 mmol	0.97 mmol	0.94 mmol	0.90 mmol
$Ni(OOCCH_3)_2·4H_2O$	0 mmol	0.03 mmol	0.06 mmol	0.10 mmol
$SbCl_3$	3 mmol	3 mmol	3 mmol	3 mmol
$Co_{1-y}Fe_ySb_3$				
Precursors	y=0	y=0.05	y=0.1	y=0.25
$Co(OOCCH_3)_2·4H_2O$	1 mmol	0.95 mmol	0.9 mmol	0.75 mmol
$FeCl_3·6H_2O$	0 mmol	0.05 mmol	0.1 mmol	0.25 mmol
$SbCl_3$	3 mmol	3 mmol	3 mmol	3 mmol

content were changed as shown in table I. After the solution was strongly agitated by argon gas bubbling at room temperature for 45 min, 11mmol sodium borohydride (NaBH₄) as a reducing agent dissolved previously in TEG was added slowly, and the Ni-doped or Fe-doped CoSb₃ compound was synthesized by heating to 513 K in an argon atmosphere, for different reaction times (1 h, 5 h and 12h). The synthesized particles in the solution were settled by centrifugation, washed with TEG and ethanol (C₂H₅OH), and then dried in an argon atmosphere. The reaction products were characterized by using a X-ray diffractometer (XRD) with Cu Kα radiation and a transmission electron microscopy (TEM).

DISCUSSION

Synthesis of Co₁₋ₓNiₓSb₃ nanoparticles

First, we investigated an influence of reaction time on the synthesis of Co₁₋ₓNiₓSb₃ nanoparticles. Figure 1 shows X-ray diffraction patterns of the powder samples (Ni content: x=0.06) synthesized for three different reaction times. For the reaction time of 1 h, the main phase detected was the antimony phase, not the skutterudite phase. For the reaction time of 5 h, the skutterudite phase was mainly detected together with the phases of Sb and MSb₂ (M=Co, Ni). However, further long reaction time (12h) brought increases of the peak intensities of Sb and

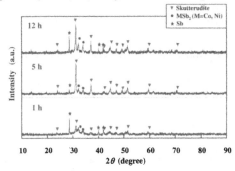

Figure 1. X-ray diffraction patterns of Co₁₋ₓNiₓSb₃ (x=0.06) nanoparticles synthesized for three different reaction times.

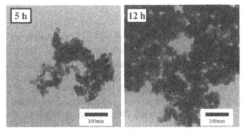

Figure 2. TEM images of nanoparticles Co₁₋ₓNiₓSb₃ (x=0.06) synthesized for different reaction time.

137

MSb$_2$ (M=Co, Ni). The skutterudite phase was apparently reduced into Sb and MSb$_2$ (M=Co, Ni) phases. From this result, the reaction time of 5 h is the best among three conditions. The reason why the skutterudite phase was reduced in the reaction time of 12 h is uncertain. Further investigation about this result would be required. TEM images of the Ni-doped CoSb$_3$ nanoparticles synthesized for the reaction time of 5 h and 12 h are shown in figure 2. Both nanoparticles synthesized for 5 h and 12 h have average size of 20-30 nm and exist as aggregates. The particle size of synthesized nanoparticles was maintained without particle growth during the synthesizing reaction.

The X-ray diffraction patterns of Co$_{1-x}$Ni$_x$Sb$_3$ nanoparticles synthesized in the four different nickel contents (x=0, 0.03, 0.06 and 0.1) for the reaction time of 5 h was measured. The skutterudite phase was mainly detected regardless of the different nickel content. However, the synthesized nanoparticles were not a single phase of skutterudite, including small amounts of Sb and MSb$_2$ (M=Co, Ni) phases. The lattice parameter of the synthesized skutterudite phase was estimated from data of the X-ray diffraction patterns of the phase. Figure 3 shows the relationship between the lattice parameter and the Ni content. The lattice parameter linearly

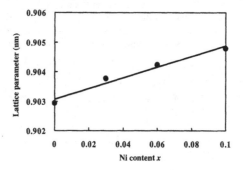

Figure 3. Relationship between lattice parameter and nickel content of Co$_{1-x}$Ni$_x$Sb$_3$ nanoparticles.

increased with increasing the Ni content, following Vegard's law. This indicates that nickel atoms are substituted into a part of cobalt sites in the crystal structure of CoSb$_3$ in proportion to Ni content. Thus, the synthesis of Co$_{1-x}$Ni$_x$Sb$_3$ nanoparticles was confirmed.

Synthesis of Co$_{1-y}$Fe$_y$Sb$_3$ nanoparticles

We also attempted to synthesize the Fe-doped CoSb$_3$ nanoparticles according to the knowledge obtained in the synthesis of Co$_{1-x}$Ni$_x$Sb$_3$ nanoparticles. The reaction time was fixed with 5 h that was the best reaction time for synthesizing Co$_{1-x}$Ni$_x$Sb$_3$ nanoparticles. Figure 4 shows the X-ray diffraction patterns of Co$_{1-y}$Fe$_y$Sb$_3$ nanoparticles synthesized in the four different nickel contents (x=0, 0.05, 0.1 and 0.25) for the reaction time of 5 h. It is found that the addition of Fe controlled the synthesis of skutterudite phase and promoted the synthesis of phases of Sb and MSb$_2$ (M=Co, Fe). Thus, we investigated again the influence of reaction time on the synthesis of Fe-doped CoSb$_3$ nanoparticle. The result of XRD analysis of the Co$_{1-y}$Fe$_y$Sb$_3$ (y=0.1) nanoparticles synthesized for the reaction time of 12 h was compared with that

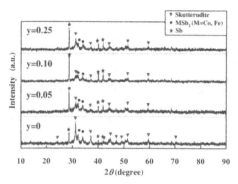

Figure 4. X-ray diffraction patterns of $Co_{1-y}Fe_ySb_3$ nanoparticles synthesized in four different Fe contents.

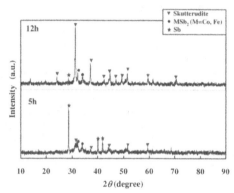

Figure 5. X-ray diffraction patterns of $Co_{1-y}Fe_ySb_3$ (y=0.1) nanoparticles synthesized for reaction time of 5 h and 12 h.

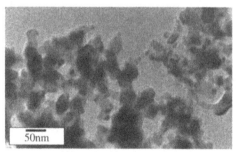

Figure 6. TEM images of nanoparticles $Co_{1-y}Fe_ySb_3$ (y=0.1) synthesized for reaction time of 12 h.

139

synthesized for the reaction time of 5 h. Figure 5 shows the X-ray diffraction patterns of powder samples (Fe content: y=0.1) synthesized for the reaction time of 5 h and 12 h. The longer reaction time (12 h) remarkably promots the synthesis of skutterudite phase. The lattice parameter estimated from the X-ray diffraction data of the skutterudite phase synthesized for the reaction time of 12 h was 0.9056 nm, which is larger than the lattice parameter (0.9029 nm) of $CoSb_3$ synthesized in this research. It can be thought that iron atoms were substituted into a part of cobalt sites in the crystal structure of $CoSb_3$. TEM image of the Fe-doped $CoSb_3$ nanoparticles synthesized for the reaction time of 12 h are shown in figure 6. Average size of the synthesized nanoparticles was 20-30 nm as well as that of Ni-doped $CoSb_3$ nanoparticles. The nanoparticles exist as aggregates. From these results, the synthesis of $Co_{1-y}Fe_ySb_3$ nanoparticles was confirmed.

CONCLUSIONS

Syntheses of the Ni-doped and Fe-doped $CoSb_3$ nanoparticles were attempted through the modified polyol process in which two main precursors of $Co(OOCH_3)_2 \cdot 4H_2O$ and $SbCl_3$ were mixed with an additive precursor of $Ni(OOCH_3)_2 \cdot 4H_2O$ or $FeCl_3 \cdot 6H_2O$ and they were reduced by $NaBH_4$ in the solvent of tetraethyleneglycol at 513 K. The reaction time and the content of additive element (Ni or Fe) were changed for synthesizing the nanoparticles. In the Ni-doped $CoSb_3$ nanoparticles, the nanoparticles synthesized for the reaction time of 5 h had the average particle size of 20-30 nm and consisted of almost the single phase of Ni-doped $CoSb_3$. The lattice parameter linearly increased with increasing the Ni content. Ni atoms were substituted into a part of Co sites in the crystal structure of $CoSb_3$ in proportion to Ni content. In the Fe-doped $CoSb_3$ nanoparticles, the nanoparticles synthesized for the reaction time of 12 h had the average particle size of 20-30 nm and consisted of almost the single phase of Fe-doped $CoSb_3$. Although the further investigation for optimizing the condition of synthesis will be required for obtaining the nanoparticles of the single phase of Ni-doped or Fe-doped $CoSb_3$, this research could clarify that the nano-size thermoelectric compound consisting of three elements can be synthesized via the modified polyol process.

REFERENCES

1. G. S. Nolas, M. Kaeser, R. T. Littleton IV, and T. M. Tritt, *Appl. Phys. Lett.* **77**, 1855 (2000).
2. H. Anno, K. Ashida, K. Matsubara, G. S. Nolas, K. Akai, M. Matsuura, and J. Nagao, *Mater. Res. Soc. Symp. Proc.* **691**, 49 (2002).
3. B. C. Sales, D. Mandrus, and R. K. Williams, *Science* **272**, 1325 (1996).
4. G. S. Nolas, J. Sharp, and H. J. Goldsmid, *Thermoelectrics: Basic Principles and New Materials Developments*, (Springer Series in Materials Science, Vol. 45), Springer, Berlin (2001), pp178-180.
5. T. Itoh, K. Ishikawa, and A. Okada, *J. Mater. Res.* **22**, 249 (2007).
6. F. Fievet, J. P. Lagier, B. Blin, B. Beaudoin, and M. Figlarz, *Sol. State Ionics* **32**, 198 (1989).
7. R. E. Cable, and R. E. Schaak, *Chem. Mater.* **17**, 6835 (2005).
8. K. Isogai, and T. Itoh, *Proc. International Symposium on EcoTopia Science 2007 (ISETS07)*, EcoTopia Science Institute, Nagoya University (2007), pp.155-157.
9. L. Yang, H.H. Hng, H. Cheng, T. Sun, and J. Ma, *Mater. Lett.* **62**, 2483 (2008).

Mater. Res. Soc. Symp. Proc. Vol. 1166 © 2009 Materials Research Society 1166-N03-17

Output Power Characteristics of Mg₂Si and the Fabrication of a Mg₂Si TE Module With a uni-leg Structure

Takashi Nemoto[1], Tsutomu Iida[2], Yohei Oguni[2], Junichi Sato[1], Atsunobu Matsumoto[2], Tatsuya Sakamoto[2], Takahiro Miyata[2], Tadao Nakajima[1] Hirohisa Taguchi[2], Keishi Nishio[2] and Yoshifumi Takanashi[2]

[1] Nippon Thermostat Co., Ltd., 6-59-2 Nakazato, Kiyose-shi, Tokyo 204-0003, Japan
[2] Department of Materials Science and Technology, Tokyo University of Science, 2641 Yamazaki, Noda-shi, Chiba 278-8510, Japan

ABSTRACT

Mg₂Si elements and a TE module with a transition metal electrode consisting of Ni were fabricated using a monobloc sintering. In order to design a structure for a thermoelectric module using Mg₂Si, we examined the correlation between the ZT values and the power-output of a single element, using Mg₂Si and Mg₂Si doped with donor impurities, such as Al and/or Bi. The observed power-outputs for single elements of Mg₂Si (ZT = 0.6), 2 at. % Bi-doped Mg₂Si (ZT = 0.65) and 1at. % Bi + 1at. % Al-doped Mg₂Si (ZT = 0.77) were 23.2 mW, 13.6mW and 19.4 mW at ΔT = 500 K (between 873 K and 373 K), respectively. Additionally, we developed and evaluated a new architecture, based on a 'uni-leg' structure Mg₂Si TE power generation module. The observed maximum values for open circuit voltage (V_{oc}) and output power (P) of the Mg₂Si TE power generation module were 109 mV and 48 mW at ΔT = 500 K (between 873 K and 373 K), respectively.

INTRODUCTION

Thermal-to-electric (TE) energy conversion from waste heat source is a viable technology that can be instrumental in improving the potential conversion efficiency of caloric power generation, such as conventional heat engines and high temperature furnaces [1]. Magnesium silicide (Mg₂Si) has been identified as a promising advanced thermoelectric material, operating in temperatures ranging from 500 to 800 K, because Mg₂Si possesses essential requisites for achieving practical use, such as abundance of its constituent elements in the earth's crust and non-toxicity of its processing-by-products [2-9]. Additionally, the dimensionless figure of merit, ZT, characterizing the efficiency of a thermoelectric material, of n-type Mg₂Si has already reached ~1.0 at 873 K [10]. However, the correlation between the ZT values and the power generation characteristics, essential to the understanding of the design of a structure for a TE power generation module, has not been sufficiently investigated.

In this experiment, three different single elements, using undoped Mg₂Si and Mg₂Si doped with donor impurities such as Al and/or Bi, were fabricated, and the open circuit voltage (V_{OC}), the output current (I) and the output power (P) of the three different elements were measured and compared. In addition this, we studied the relation of Seebeck coefficient and thermal conductivity to the value of V_{OC} and the relation of electrical conductivity to the value of I. Additionally, as a feasible TE module, we developed a pin-fin structure TE module, which is the new architecture based on the uni-leg structure TE module, using an undoped Mg₂Si source. This structure offers concise assembly and better radiation performance at cool-side to maintain a temperature gradient, together with a fail-safe composition for the anticipated thermal expansion at a temperature of operation of ~870 K. The output power value was evaluated for this module.

EXPERIMENT

As a starting material for the fabrication of the single elements, pre-synthesized polycrystalline un-doped Mg_2Si and Mg_2Si doped with donor impurities such as Al and/or Bi, fabricated by UNION MATERIALS INC, were used. Mg_2Si sources used in this experiment are listed in Table.1. The materials were sintered using a plasma-activated sintering (PAS) technique, and, at the same time, Ni electrodes were formed on the Mg_2Si by employing a monobloc PAS technique, performed at 1048 K for 2 min at a pressure of 29.4 MPa in an Ar (0.06 MPa) atmosphere. The size of the sintered samples was 15 mm in diameter and 10 mm in thickness. The measured single elements were 2x2x10 mm³. The pin-fin structure TE module using un-doped Mg_2Si had a 4.3 x 4.3 x 3mm Mg_2Si base, over which 4 square Mg_2Si pins, which were 7mm tall and had a square cross section of 2 x 2 mm, were arranged in a regular, in-line pattern. The pictures of a single element and the pin-fin structure TE module are shown in Fig.1. The open-circuit voltages and the thermoelectric power outputs under a temperature differences ΔT, ranging from 100 to 500 K, were measured using a UNION MATERIAL UMTE-1000M. The top of an element or a module was heated with electrical heater at 473 K ~873 K, whereas their bases were maintained at 373 K, by using a water-cooled aluminum block. The output current and output power were measured under closed circuit conditions. As an external load, an element, identical to the measured element, was used. The schematic diagram of the measurement method is shown in Fig.1.(C). The Seebeck coefficient (S) and the electrical conductivity (σ) of the sintered samples were measured over the temperature range from 350 to 860 K using ULVAC-RIKO ZEM-2 equipment. Measured S_{Ave} and σ_{Ave} values as a function of the temperature difference ΔT ranging from 100 to 500 K are listed in Table.2. S_{Ave} and σ_{Ave} correspond to an averaging of the $S_{(T)}$ and $\sigma_{(T)}$ values through the entire temperature difference ΔT. The thermal conductivity was measured over a temperature from 300 to 873 K, employing the Laser flash method, using ULVAC-RIKO TC-7000H.

Table.1. List of Mg_2Si sources.

Sample name	Al doping (at.%)	Bi doping (at.%)	Union Materials Product ID
MS	0	0	MGSI-SG-Un
MS-BI2	0	2	MGSI-SG-BI2
MS-BI1AL1	1	1	MGSI-SG-BI1AL1

  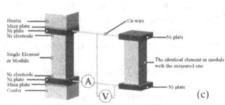

Fig.1. (a) : Single element of Mg_2Si with Ni electrodes. (b) : Pin-fin structure TE module. (c) : The schematic diagram of the measurement method.

Table.2. The values of S_{Ave} and σ_{Ave}

Sample name	MS		MS-BI2		MS-AL1BI1	
ΔT	S_{Ave} (μV/K)	σ_{Ave} (S/m)	S_{Ave} (μV/K)	σ_{Ave} (S/m)	S_{Ave} (μV/K)	σ_{Ave} (S/m)
100	-147.5	1.0×10^5	-112.6	1.3×10^5	-111.0	1.6×10^5
200	-177.2	7.9×10^4	-125.0	1.2×10^5	-122.0	1.4×10^5
300	-207.4	6.3×10^4	-135.8	1.1×10^5	-132.8	1.3×10^5
400	-221.5	5.6×10^4	-144.6	1.1×10^5	-142.6	1.2×10^5
500	-226.7	5.3×10^4	-152.3	9.9×10^4	-151.3	1.0×10^5

RESULTS & DISCUSSION

Fig.2 shows the temperature dependence of the dimensionless figure of merit, ZT, which is an index of thermoelectric performance and is closely associated with the thermal-to-electric conversion efficiency. The ZT value for TE material is given by the following equation:

$$ZT = \frac{S^2\sigma}{k}T , \qquad (1)$$

where S is the Seebeck coefficient, σ is the electrical conductivity and κ is the thermal conductivity. The observed ZT values for MS, MS-BI2 and MS-AL1BI1 were 0.6, 0.65 and 0.77, respectively, at 860 K. The measured and calculated open-circuit voltage (V_{OC} and $V_{OC(cal)}$) values of elements using these materials under a temperature gradient, ΔT, ranging from 100 to 500 K are shown in Fig.4. The $V_{OC(cal)}$ value is defined as follows:

$$V_{OC(cal)} = \int_{T_C}^{T_H} S(T)dT , \qquad (2)$$

where T_H is a hot-side temperature and T_C is a cold-side temperature. The highest observed V_{OC} values for MS, MS-BI2 and MS-AL1BI1 were 95.3 mV, 61.5 mV and 66.8 mV, respectively, at $\Delta T = 500$ K (between 873 K and 373 K). The observed V_{OC} values for MS were higher than those of MS-BI2 and MS-AL1BI1 because the Seebeck coefficient of MS is greater than that of MS-BI2 and MS-AL1BI1, as shown in Table.2. The measured V_{OC} values for all samples were

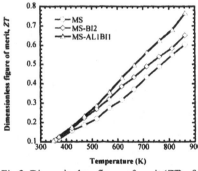

Fig.2. Dimensionless figure of merit (ZT) of sintered samples in the range from 350 to 860 K

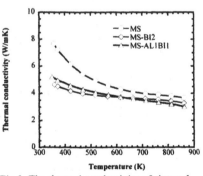

Fig.3. The thermal conductivity of sintered samples in the range from 350 to 860 K.

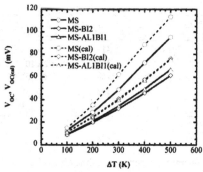

Fig.4. Measured and calculated open-circuit voltages of single elements as a function of the temperature difference ΔT ranging from 100 to 500 K.

Table.3. The values of V_{OC} / $V_{OC(cal)}$

ΔT	MS	MS-BI2	MS-AL1BI1
100	0.87	0.80	0.85
200	0.82	0.79	0.86
300	0.79	0.79	0.86
400	0.82	0.80	0.87
500	0.84	0.81	0.88

lower than those calculated. The ratios of V_{OC} to $V_{OC(cal)}$ were calculated in order to confirm the influence of the thermal conductivity on the V_{OC} values and these are listed in Table.3. Although the thermal conductivity value of MS was higher than those of MS-BI2 and MS-AL1BI1 as shown in Fig3, the obtained $V_{OC}/V_{OC(cal)}$ value for MS was comparable to those of MS-BI2 and MS-AL1BI1. It may be that the influence in other heat transfer, such as a radiation from heat source, was greater than the influence of a heat transfer through the element.

Fig.5 shows the measured output current (I) values as a function of the temperature difference ΔT ranging from 100 to 500 K. The highest I values with an external load for MS, MS-BI2 and MS-AL1BI1 were 448 mA, 393 mA and 430.5 mA, respectively, at ΔT = 500 K. Although, as shown in Table.2, the σ_{Ave} values for MS-BI2 and MS-AL1BI1 were higher than that for MS, the observed I values of MS-BI2 and MS-AL1BI1 were slightly lower than that of MS. The reason why the I values for MS-BI2 and MS-ALBI1 were lower than that for MS is that the output voltage values with an external load of MS-BI2 and MS-ALBI1 were lower than that of MS, caused by the low S_{Ave} values of MS-BI2 and MS-ALBI1. As an additional reason, it seems that the influence of the loss of current in the measured circuit caused by an wiring and contact resistance for MS-BI2 and MS-AL1BI1 was relatively larger than that for MS because the internal resistance values of MS-BI2 and MS-AL1BI1 were lower than that of MS.

The results of the output power (P) measurements with load resistance for all samples under a temperature gradient ΔT ranging from 100 to 500 K are shown in Fig.6. The highest observed P values for elements of MS, MS-BI2 and MS-AL1BI1 were 23.2 mW, 13.6mW and

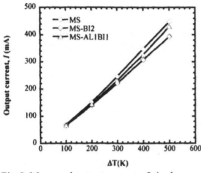

Fig.5. Measured output current of single elements as a function of the temperature difference ΔT ranging from 100 to 500 K.

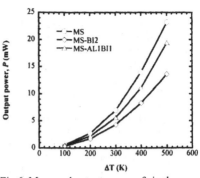

Fig.6. Measured output power of single elements as a function of the temperature difference ΔT ranging from 100 to 500 K.

19.4 mW, respectively, at $\Delta T = 500$ K. It was noted that the output power value of MS was greater than that of MS-BI2 and MS-AL1BI1, although the ZT value for MS was lower than that for MS-BI2 and MS-AL1BI1. It was revealed that if the circuit of a module that suits the thermoelectric characteristic of material is not properly prepared, even if material with high value of ZT is used, the performance could not be adequately demonstrated.

As a feasible Mg₂Si TE device, as shown in Fig.1. (b), we have developed a pin-fin structure module with an n-type uni-leg Mg₂Si element using the MS material. This device structure offers concise assembly and better radiation performance at the cool-side to maintain a temperature gradient, together with a fail-safe composition for the anticipated thermal expansion at a temperature of operation of ~870 K. The results of the P measurement for this module with a external load are shown in Fig. 7. The maximum P and V_{OC} values for this TE module were 48 mW and 109 mV, respectively, at $\Delta T = 500$ K. The calculated ratio of V_{OC} to $V_{OC(cal)}$ was 0.96 at

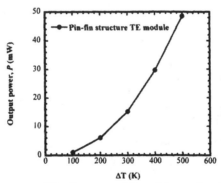

Fig.7. Measured output power of Pin-fin structure TE module as a function of the temperature difference ΔT ranging from 100 to 500 K.

145

ΔT = 500 K, this value is higher than that for single element of MS. Additionally, we have confirmed that the open circuit voltage and out put power for the pin fin structure module were larger than values obtained from the module comprising four single elements, connected in parallel. It seems reasonable to conclude that the pin-fin structure offers better radiation performance at the cool-side to maintain the temperature gradient.

CONCLUSIONS

Mg_2Si elements and a TE module with a transition metal electrode consisting of Ni were fabricated using monobloc sintering and the thermoelectric powers were evaluated. A regular relation was not observed between the ZT values and the output power values. In the case of practical use, it seems that the thermal influences, such as heat radiation from the heat source, and electrical influences, such as resistance of wiring and contacts, were large. Therefore, if a low resistance element is used, it is required that resistance of wiring and contacts in the TE module should be lowered in order to effectively demonstrate the performance of the material. For the pin-fin structure TE module, this structure improves the radiation performance at cool-side to maintain a temperature gradient. The obtained P values for MS, MS-BI2 and MS-AL1BI1 were 23.2 mW, 13.6mW and 19.4 mW, respectively, at ΔT = 500 K. The maximum P values of this the pin-fin structure TE module was 48 mW at ΔT = 500 K.

ACKNOWLEDGMENTS
This work was partly supported by a Grant-in-Aid for Research (B) by the Japanese Ministry of Education, Science, Sports, and Culture.

REFERENCES

1. Terry M. Tritt and M.A. Subramanian, MRS Bull. 31, 188 (2006)
2. R.G. Morris, R.D. Redin and G.C. Danielson, *Phys. Rev.* **109**, 1909 (1958).
3. V.E Borisenko, *Semiconducting Silicides, Springer, Berlin*, 285 (2000).
4. Y. Noda, H. Kon, Y. Furukawa, N. Otsuka, I.A. Nishida, K. Masumoto, *Mater. Trans. JIM* **33**, 845 (1992).
5. Y. Noda, H. Kon, Y. Furukawa, N. Otsuka, I.A. Nishida, K. Masumoto, *Mater. Trans. JIM* **33**, 851 (1992).
6. M. Akasaka, T. Iida, T. Nemoto, J. Soga, J. Sato, K.Makino, M. Fukano and Y. Takanashi, *Journal of Crystal Growth*, **304**, 196 (2007).
7. M. Akasaka, T. Iida, A. Matsunoto, K. Ymanaka, Y. Takanashi, T. Imai and N. Hamada, *Journal of Applied Physics*, **104**, 013703 (2008)
8. J. Tani, H. Kido, *Physica*, **B364**, 218 (2005).
9. I. Aoyama, H. Kaibe, L. Rauscher, T. Kanda, M. Mukoujima, S. Sano and T. Tsuji, *Japanese Journal of Applied Physics*, **44**, 4275 (2005)
10. M. Fukano, T. Iida, K. Makino, M. Akasaka, Y. Oguni and Y. Takanashi in *Crystal Growth of Mg2Si by the Vertical Bridgman Method and the Doping Effect of Bi and Al on Thermoelectric Characteristics,* edited by T.P. Hogan, J. Yang, R. Funahashi and T.M. Tritt (Mater. Res. Soc. Symp. Proc. **1044**, 2008) pp.247-252

Mater. Res. Soc. Symp. Proc. Vol. 1166 © 2009 Materials Research Society 1166-N03-21

Thermoelectric Properties of Sb-Doped Sintered Mg$_2$Si Fabricated Using Commercial Polycrystalline Sources

Naoki Fukushima[1], Tsutomu Iida[1], Masayasu Akasaka[2], Takashi Nemoto[3], Tatsuya Sakamoto[1], Ryo Kobayashi[1], Hirohisa Taguchi[1], Keishi Nishio[1] and Yoshifumi Takanashi[1]

[1]Department of Materials Science and Technology, Tokyo University of Science, 2641 Yamazaki, Noda-shi, Chiba 278-8510, Japan
[2]Dow Corning Toray Co., Ltd., 2-2 Chigusa-Kaigan Ichihara City, Chiba 299-0108, Japan
[3]Nippon Thermostat Co., Ltd., 6-59-2 Nakazato, Kiyose-shi, Tokyo 204-0003, Japan

ABSTRACT

The thermoelectric (TE) properties, such as the Seebeck coefficient, the electrical and thermal conductivities, and the output power, of Sb-doped n-type Mg$_2$Si were studied. A commercial polycrystalline source was used for the source material for the Mg$_2$Si. TE elements with Ni electrodes were fabricated by using a monobloc plasma-activated sintering (PAS) technique. Compared with undoped samples, the ZT values of the Sb-doped samples were higher over the whole temperature range in which measurements were made; the maximum value for the Sb doped Mg$_2$Si was 0.72 at 864 K. The TE characteristics of Sb-doped samples were found to be comparable to those of Bi-doped ones, and no significant difference in ZT value was observed between them. Provisional results showed that the maximum value of the output power was 6.75 mW for the undoped sample, 4.55 mW for a 0.5 at% Sb doped sample, and 5.25 mW for a 1 at% Sb doped sample with $\Delta T = 500$ K (between 873 K and 373 K).

INTRODUCTION

Magnesium silicide (Mg$_2$Si) has been identified as a well-balanced thermoelectric material that covers the temperature range from 500 to 800 K[1-2], and one that fits well with the operating temperatures of industrial furnaces, automobile exhausts, and incinerators. The important features of Mg$_2$Si include the fact that it has been identified as an environmentally-benign material, that its constituent elements are abundant in the earth's crust, and that it is non-toxic[3-5].

Although thermoelectric devices have obvious merits in terms of power generation, one of the reasons why thermoelectric devices are not more widely used at present is that the cost-per-watt of thermoelectric power generation has been too high to allow it to displace existing technologies. Mg$_2$Si has significant advantages for various applications in terms of both lower raw materials costs due to the abundance of its constituent elements, and its lighter weight compared with conventional thermoelectric materials such as Bi$_2$Te$_3$ and CoSb$_3$.

In order to optimize the thermoelectric properties of Mg$_2$Si for practical applications, it needs to be doped, and typical device operating temperatures require that any substitutional element used as a dopant be highly stable. Stable dopant elements in Mg$_2$Si are needed to ensure long lifetime operation at elevated temperatures. Aluminum (Al), bismuth (Bi) and antimony (Sb) are well known n-type dopants for Mg$_2$Si. Calculations from first principles show that Sb is the most stable in Mg$_2$Si compared with the other dopants[6]. The doping characteristics of Sb for Mg$_2$Si crystals doped during growth or for Mg$_2$Si sintered directly from Mg and Si sources by solid-phase diffusion have been reported in earlier studies. [7-8] In this paper, we report on the characteristics of Sb in sintered Mg$_2$Si from a commercial polycrystalline Mg$_2$Si source.

EXPERIMENTAL

Pre-synthesized commercial poly-crystalline Mg$_2$Si sources were provided by UNION MATERIAL with Sb doping of 0.5, 1.0, 2.0, and 3.0 at.%, and Bi doping of 0.5, 1.0 and 2.0 at.%. Doping was carried out during synthesis of polycrystalline Mg$_2$Si by UNION MATERIAL. The source materials were pulverized to powder with sizes of 75 µm or less and then sintered by a Plasma Activated Sintering (PAS) technique using an ELENIX PAS-III-Es. The Mg$_2$Si powder was placed in a graphite die and sintered at 1123 K for 10min at a pressure of 29.4 MPa, with a heating rate of 100 K/min. The samples were then cut using a wire saw. X-ray powder diffraction (XRD) analysis of these samples was performed using CuK$_\alpha$ radiation. The thermoelectric properties, including the Seebeck coefficient, S, and the electrical conductivity, σ, were measured over a temperature range from room temperature to 850 K using ULVAC-RIKO ZEM-2 equipment. A laser flash method, using an ULVAC-RIKO TC-7000H, was employed to measure the thermal conductivity, κ, over a temperature range from room temperature to 850 K; the dimensionless figure of merit, ZT, was then estimated. The Hall carrier concentration and Hall mobility were measured at room temperature using the van der Pauw method. The magnetic flux density was 0.53 T, and the current was 100 mA. The output power, with the temperature gradient, ΔT, ranging from 100 to 500 K, was measured using a UNION MATERIAL UMTE-1000M. For the samples used for output power measurements, Ni electrodes were formed by means of a monobloc sintering process in the temperature range from 1093 K to 1123 K. The sample size for the output power measurements was 2x2x10 mm^3.

RESULTS AND DISCUSSION

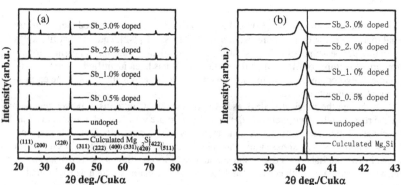

Figure 1.(a) X-ray powder diffraction patterns of the commercial Mg$_2$Si source provided by UNION MATERIAL, (b) (220) X-ray diffraction patterns of Mg$_2$Si.

Fig. 1 shows typical XRD patterns of the polycrystalline Mg_2Si sources provided by UNION MATERIAL for undoped and various Sb concentrations. For all specimens clear XRD peaks corresponding to Mg_2Si were observed as shown in Fig.1 (a), indicating neither residual Mg metal in the source nor formation of other Sb-related compounds, such as Mg_3Sb_2, even at higher Sb content. In terms of device degradation at the expected elevated operating temperature, elimination of metallic Mg in the processed sample is crucial; thus, the Mg_2Si source material from UNION MATERIAL meets this requirement. Fig.1(b) shows an enlarged XRD pattern of the dominant (220) diffraction peaks. Compared with the undoped sample, the observed diffraction peaks of the Sb doped samples shift toward a lower angle and their widths broaden with increasing Sb content due to the changing lattice constant.

The samples fabricated in this experiment are listed in Table I, with the doping conditions and the results of Hall measurements. In order to compare the behavior of Sb as a donor, Bi doped samples in a comparable concentration range were also prepared using the same PAS process. The undoped sample and those with 0.5 and 1 at% Sb were fabricated successfully; however, some samples doped with 2 and 3 at% Sb contained fissures. For the PAS sintered samples, XRD peaks associated only with Mg_2Si were observed, which were as observed for the source material, while the same peak shift and broadening were also observed with increasing Sb content. As for the samples with 2 and 3 at% Sb, a sawing machine was used to form small pieces to fit the ZEM-2 and for laser flash measurements; however, this was insufficient to enable measurement of the thermoelectric properties owing to fissures or cracks. The friable condition of samples with 2 and 3 at% Sb were reproducible, although the reason for this is still unclear.

Table I. Fabricated samples with the results of Hall measurements.

Sample	Dopant content (at%)	Hall Carrier concentration (cm^{-3})	Mobility (cm/Vs)	Product No
undoped	undoped	4.28×10^{19}	145	MGSI-MG-UN
Sb_0.5	0.5	1.62×10^{20}	88.7	MGSI-MG-SB05
Sb_1.0	1	2.58×10^{20}	39.1	MGSI-MG-SB1
Sb_2.0	2	------	------	MGSI-MG-SB2
Sb_3.0	3	------	------	MGSI-MG-SB3
Bi_0.5	0.5	1.10×10^{20}	69.5	MGSI-MG-BI05
Bi_1.0	1	1.45×10^{20}	81.9	MGSI-MG-BI1
Bi_2.0	2	1.36×10^{20}	74.8	MGSI-MG-BI2

Hall measurements indicate that the carrier concentration of the Sb doped samples increases with increasing Sb content. It is seen that the peak electrical activation, with a carrier concentration of $\sim 2 \times 10^{20}$ cm^{-3}, is obtained with 1 at% Sb initial doping. X.S. Lin et al have reported that Mg_2Si doped with 18 at% Sb resulted in a relatively small carrier concentration, 1.6×10^{20} cm^{-3}[8], perhaps indicating that the amount of Sb was beyond the solid solubility in Mg_2Si. These authors speculated that this low carrier concentration, even at such a high doping level, might be associated with Mg vacancies, which act as acceptors, giving rise to a phenomenon similar to compensation[8]. In our present case, it seems that an acceptable carrier concentration for TE device operation is obtained with Sb-doped Mg_2Si with a content between 0.5 and 1 at%.

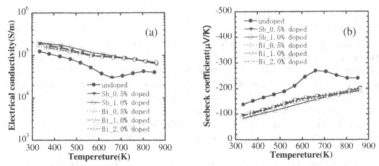

Figure 2. Temperature dependent (a) Seebeck coefficient, and (b) electrical conductivity of sintered samples, over the temperature range from 350K to 860K.

Results of the Seebeck coefficient and the electrical conductivity of the sintered samples are shown Fig. 2 as a function of sample temperature. The observed Seebeck coefficient was negative for all samples, indicating n-type conductivity. Compared with the undoped sample, an increase in the electrical conductivity and a decrease in the absolute value of the Seebeck coefficient were observed with increasing Sb. The observed difference between the samples Sb_0.5 and Sb_1.0 was rather small. It is noted that the values of S and σ of the Sb doped samples are comparable to those of the Bi doped samples as a function of temperature.

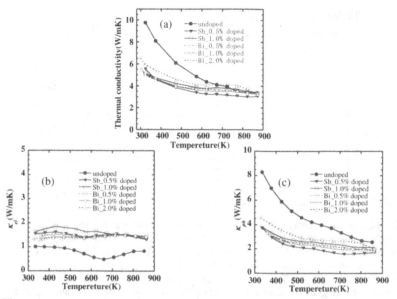

Figure 3. The thermal conductivity of sintered samples in the range from 350 to 860 K (a) total κ, (b) electronic component κ_{el}, and (c) lattice component κ_{ph}.

Variations in the thermal conductivity, κ, of Sb- and Bi-doped specimens for temperatures ranging from room temperature to 873 K are shown in Fig.3 (a). The observed κ values decrease monotonically with increasing sample temperature up to ~723 K, indicating an increase in phonon scattering, and start to level out to 3 to 4 W/mK beyond ~600 K. Regarding Fig. 3 (b) and (c), the observed thermal conductivity, κ, is separated into contributions from the lattice (κ_{ph}) and the electronic (κ_{el}) components. Each of the κ_{el} and κ_{ph} values was calculated using the Wiedeman-Franz law, $\kappa_{el} = L_0 \sigma T$, where L_0 is the Lorentz number, 2.45×10^{-8} V^2/K^2 [8]. Incorporation of Sb in Mg_2Si brought about a decrease in the κ_{ph} values over the whole temperature range, and exhibited comparable temperature dependences to those of Bi-doped samples. At about 800 K, the observed κ_{ph} for doped and undoped specimens are similar. Since the thermal conductivity of Sb doped Mg_2Si behaves in a similar manner to Bi doped Mg_2Si and thus will maintain a similar temperature gradient for practical device operation at elevated temperatures, and since it is less toxic than Bi, Sb is a good substitute for Bi.

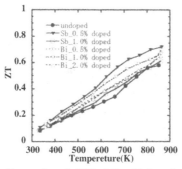

Figure 4. Dimensionless figure of merit (ZT) of Sb-, Bi-doped, and undoped samples.

Fig. 4 shows the evaluated values of ZT for Sb- and Bi-doped samples together with data for an undoped one. Compared with the ZT value of the undoped sample, Sb-doped samples have higher values over the whole temperature range. The maximum values of Sb doped Mg_2Si are 0.72 at 864 K (Sb_0.5) and 0.61 at 861 K (Sb_1.0). It can be seen that the characteristics of the Sb-doped samples are comparable to those of the Bi-doped ones, indicating that Sb is a good alternative to Bi as an n-type dopant in Mg_2Si.

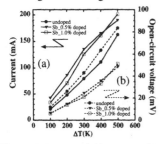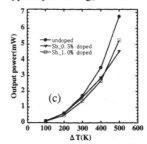

Figure 5. Measured (a) current, (b) open circuit voltage, and (c) output power for undoped and Sb-doped specimens as a function of temperature difference, ΔT, ranging from 100 to 500 K.

Fig.5 shows the results of current, open circuit voltage, and output power measurements for the Sb-doped and undoped samples as a function of temperature. The top part of the 2x2x10 mm^3 rectangular parallelepiped sample was heated from 473 up to 873 K, while the bottom part was maintained at 373 K. The measured open circuit voltages for the Sb-doped samples were lower than that of the undoped one. For the output power measurement, the voltage and current were measured under closed circuit conditions by means of an external load consisting of a sample with the same dimensions as the measured one. The samples used for output power measurement were sintered at the slightly lower temperature of 1093 K because of leakage of Ni metal during the monobloc PAS sintering process. However, the thermoelectric properties, such as the value of ZT, of Mg$_2$Si processed at 1093 K are comparable to those of samples sintered at 1123 K. It seems that the Sb doping causes the output current to increase and the voltage to drop. Provisional results obtained show the maximum values of the output power for $\Delta T = 500$ K (between 873 K and 373 K) were 6.75 mW (undoped), 4.55 mW (Sb_0.5), and 5.25 mW (Sb_1.0).

CONCLUSION

Sb-doped n-type Mg$_2$Si was studied using commercial Mg$_2$Si source material. The Seebeck coefficient, and the electrical and thermal conductivities of plasma-sintered samples were measured. For output power measurements, Ni electrodes were formed using a monobloc plasma-activated sintering (PAS) technique. With regard to the Mg$_2$Si source material produced by UNION MATERIAL, incorporation and electrical activation of Sb were sufficient to realize TE samples. The temperature dependent behavior of the thermal conductivity with Sb doping was comparable to that obtained with Bi doping. Therefore, Sb can be used in place of Bi as an n-type dopant in Mg$_2$Si.

ACKNOWLEDGMENTS

This work was partly supported by a Grant-in-Aid for Research (B) by the Japanese Ministry of Education, Science, Sports, and Culture.

REFERENCES

[1]Victor E.Boriseneko, *Semiconducting Silicide*, (Springer, Berlin, 2000), p.285.

[2]R.G.Morris, R.D.Redin and G.C.Danielson , *Phys. Rev.* **109,** 1909 (1958)

[3]S.Bose, H.N.Acharya, and H.D.Banerjee, J.Mater. Sci. 28 (1993) 5461.

[4]Y.Noda, H.Kon, Y.Furukawa, N.Otuka, I.A.Nishida, K.Masumoto. *Mater. Trans.* JIM. **33,** 845 (1992).

[5]M.Akasaka, T.Iida, T.Nemoto, J.Soga, J.Sato, K.Makino, M.Fukano, and Y.Takanashi, *Journal of Crystal Growth.* **304,** 196 (2007).

[6]A.Kato, unpublish.

[7]J.Tani, H.Kido, Intermetallics, **15,** 1202 (2007).

[8]X.S.Lin, D.Wang, M.Beekman, and G.S. Nolas, *MRS Stmp. Proc,* **1044** (2008).

[9]C.Kittel, *Introduction to Solid State Physics.* 8th ed. (Wiley, NJ, 2005), pp.156-157

Mater. Res. Soc. Symp. Proc. Vol. 1166 © 2009 Materials Research Society   1166-N03-26

Thermoelectric Properties and Microstructure of Large-Grain Mg₂Sn Doped With Ag

H. Y. Chen and N. Savvides
CSIRO Materials Science and Engineering, Sydney 2070, Australia

ABSTRACT

Mg₂Sn ingots, doped p-type by the addition of 0–1.0 at. % Ag, were prepared by the vertical Bridgman method at growth rates ~ 0.1 mm/min. The crystalline quality and microstructure of ingots were analyzed by x-ray diffraction, scanning electron microscopy and energy-dispersive X-ray spectroscopy. The single-phase Mg₂Sn ingots consist of highly oriented large grains. Measurements of the Hall coefficient, Seebeck coefficient α, and electrical conductivity σ in the temperature range 80–700 K were conducted to study the dependence on the silver content, and to determine the thermoelectric power factor $\alpha^2 \sigma$ which reached a maximum value 2.4×10^{-3} W m^{-1} K^{-2} at 410 K for 1.0 at.% Ag content.

INTRODUCTION

Thermoelectric materials convert heat energy directly to electricity or electrical energy to refrigeration. Their conversion efficiency is determined by the figure of merit $ZT = \alpha^2 \sigma / \kappa$ where α is the Seebeck coefficient, σ is the electrical conductivity, and κ is the thermal conductivity. Recent world-wide advances in thermoelectrics [1-3] have built on progress in the 1950s–1980s [4-7].

The commercial viability for thermoelectric power generation depends not only on the efficiency but also on the cost of materials and their toxicity. The compounds Mg₂X (X = Si, Ge, Sn) and their solid solutions are promising candidates to satisfy these requirements [8-10] and also have low weight to power ratio. The high vapor pressure and chemical reactivity of Mg, however, has delayed the development of these compounds for thermoelectric applications. Recent work by Zaitsev et al. [11] has shown that $ZT \approx 1.1$ can be achieved in Mg₂Si₁₋ₓSnₓ solid solutions prepared by direct co-melting. Mechanical alloying [12], solid-state reaction [13], spark plasma sintering [14], and hot pressing [15, 16] have also been used.

We have undertaken a detailed investigation of the crystal growth and thermoelectric properties of the Mg₂X compounds [17, 18]. In this paper we report on the preparation and microstructure of Mg₂Sn crystalline ingots grown from the melt using the Bridgman technique. The ingots were doped with silver (0–1.0 at. % Ag) to obtain p-type material, and their thermoelectric properties were studied as a function of Ag content.

EXPERIMENT

The investigated Mg₂Sn ingots were grown from the melt using a vertical resistive tube furnace. High-purity Mg (4N), Sn (6N), and Ag (3N) were mixed with the desired atom ratio (Table 1), then placed into a high-purity boron nitride crucible (with screwed cover) and sealed in a quartz ampoule under a mixture of Ar and H₂ gases at 0.8 MPa. The sealed quartz ampoule was suspended in the furnace which has a sharp temperature gradient toward the bottom. Initially the ampoule was held at 1093 K for 1 hour, then lowered through the thermal gradient at a fixed velocity of 0.1 mm/min to achieve directional solidification. Thereafter the quartz ampoule was cooled to 573 K, at a rate of 1 K/min, where it was held for 5 hours.

Table 1. Sample ID, Mg/Sn atom ratio, and Ag content of starting materials

	Ingots	
Sample	Mg: Sn	at. % Ag
1	2.00: 1	0
2	2.03: 1	0.05
3	2.03: 1	0.15
4	2.05: 1	0.25
5	2.05: 1	0.50
6	2.05: 1	1.00

The high vapour pressure and extreme reactivity of Mg at high temperatures leads to loss of Mg from the melt and results in Sn-rich ingots. To compensate for this loss the ingots were prepared with a small excess of Mg, 3–5 at. %. Doping with silver also presents problems. Although silver is intended as a substitutional dopant it can also form alloys with Mg and Sn which tend to segregate toward the top of the ingot. To circumvent these problems the samples for study were cut from the bottom part of the ingots. The crystalline structure of the samples was investigated using a Philips X'Pert PRO x-ray diffractometer with Cu K radiation. The microstructure was studied using a JEOL JSM-5400 field emission scanning electron microscope (SEM) fitted with an energy-dispersive x-ray (EDX) spectrometer. Measurements of α and σ used samples about $2\times3\times(12-15)$ mm^3, while Hall measurements (sample size $1\times4\times8$ mm^3) were done at a dc field of 3 T using a Physical Properties Measurement System (PPMS, Quantum Design Inc.).

RESULTS

Figure 1(a) shows x-ray diffraction data of coarse powder derived from the middle section of the ingots. The x-ray peaks are indexed to the anti-fluorite (space group $Fm3m$) crystal structure. It is seen that all the samples are single-phase Mg$_2$Sn except for small amounts of free Sn in samples 1 and 6. Also, all samples show strong preferential crystallographic orientation, as illustrated by Figure 1(b) which shows data from three different surfaces of specimens cut from ingot 6.

154

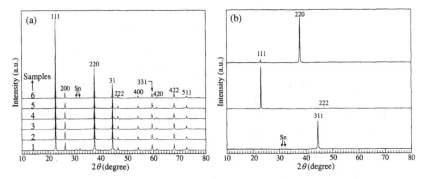

Figure 1. X-ray diffraction of (a) coarse powder derived from the six ingots, as a function of Ag content, and (b) three different surfaces of specimens cut from ingot 6 showing preferential orientation consistent with large aligned polycrystals.

Figure 2 shows the room-temperature dependence of the lattice constant a on Ag-content and hole concentration p (calculated from the Hall measurement). Both parameters increase with the initial addition of Ag but reach nearly constant values at about 0.50 at. % Ag content. This suggests that Ag-doping may be saturating at about 6×10^{19} cm^{-3}, and a large excess of Ag may exist as precipitates of Mg-Ag or Mg-Ag-Sn alloys.

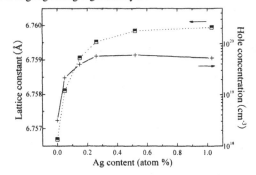

Figure 2. Effect of Ag content on the lattice constant and hole concentration.

Figure 3 shows SEM micrographs of selected areas of samples chosen for semi-quantitative compositional analysis by EDX. Samples with 0–0.25 at. % Ag content (images a and b) are single-phase Mg$_2$Sn except for small precipitates of free Sn and the occasional pore or cavity. Higher Ag-content leads to precipitates of Mg-Ag alloy and free Sn (images c and d). These particles are not uniformly distributed and their concentration is as high as 3 per 100 μm field of view.

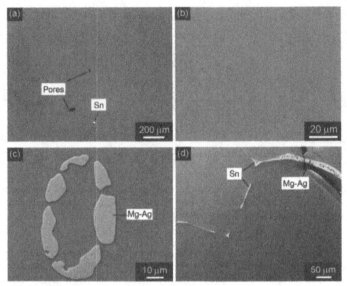

Figure 3. SEM micrographs (a, b) of samples with 0–0.25 at. % Ag content showing single phase Mg₂Sn with the occasional free Sn and pores; and (c, d) of samples with 0.5–1.0 at. % Ag content showing single phase Mg₂Sn and small precipitates of Mg-Ag alloy and free Sn.

Measurements of the Seebeck coefficient α are given in Figure 4. Sample 1 (un-doped) shows a transition from p-type conduction at low temperatures to n-type conduction above 250 K. The peak in α below 200 K is characteristic of lightly doped semiconductors where the high thermal conductivity leads to a phonon drag contribution to the Seebeck coefficient. Samples 2–6, containing increasing amount of Ag dopant, show p-type behaviour in the whole temperature range of measurement: α increases almost linearly with T to a broad peak, then decreases at higher T as n-type carriers are thermally excited to the conduction band. Characteristically, doped samples have broad Seebeck peaks (150–210 μV K^{-1} at T = 350–550 K) that shift to higher T as the hole concentration increases.

The room temperature electrical conductivity (Figure 5) increases by two orders when silver is added to the Mg₂Sn compound. At the lower temperatures phonon-electron and electron-electron scattering in the doped samples lead to $\sigma \sim 1/T$ dependence. However, for doped samples the onset of intrinsic conduction at about 400 K causes a slight rise in carrier concentration which leads to the conductivity beginning to rise with increasing temperature.

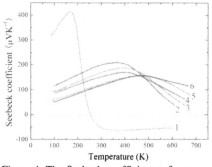

Figure 4. The Seebeck coefficient α for un-doped and Ag-doped samples.

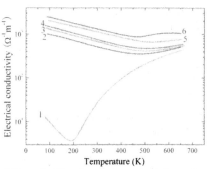

Figure 5. The electrical conductivity σ for un-doped and Ag-doped samples.

Figure 6 shows values for the thermoelectric power factor, $P = \alpha^2 \sigma$, which reaches a broad peak P_{max} at 300–450 K for each of the doped samples. This peak is evident even when the Ag-doping is very low (e.g., 0.05 at. % Ag in sample 2). As expected, sample 6, cut from the most heavily doped ingot, has the highest power factor, $P_{max} = 2.4 \times 10^{-3}$ W m^{-1} K^{-1} at 410 K. This is comparable to the best bulk thermoelectric material so far, nano-structured Bi$_2$Te$_3$ which has $P_{max} = (2–4.5) \times 10^{-3}$ W m^{-1} K^{-2} at 298–623 K leading to $ZT = 0.8$–1.4 [3]. Thermal conductivity measurement on sample 6 using PPMS [19] yield a figure of merit 0.16 at 400 K.

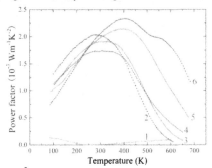

Figure 6. The power factor $P = \alpha^2 \sigma$ for undoped and Ag-doped samples.

CONCLUSIONS

Ingots of Mg$_2$Sn, doped with 0–1 at. % Ag, were grown from the melt by the Bridgeman technique. X-ray diffraction and SEM/EDX element analysis typically show the ingots are large-grain polycrystals with preferential orientation along the growth direction. The slightly-doped samples (0.05–0.25 at. % Ag) are free of secondary phases while samples containing $\geq$ 0.5 at. % Ag show small precipitates of Mg-Ag alloy and free Sn. The doped samples have high electrical conductivity and broad peaks in the Seebeck coefficient (150–210 $\mu V\ K^{-1}$) at $T = 350$–550 K. Here the power factor is optimum and reaches a maximum value $2.4 \times 10^{-3}\ W\ m^{-1}\ K^{-2}$ in samples with 1.0 at. % Ag-content.

REFERENCES

1. R. Venkatasubramanian, E. Siivola, T. Colpitts, and B. O'Quinn, Nature **413**, 597 (2001).
2. K.F. Hsu, S. Loo, F. Guo, W. Chen, J.S. Dyck, C. Uher, T. Hogan, E.K. Polychroniadis, and M.G. Kanatzidis, Science **303**, 818 (2004).
3. B. Poudel, Q. Hao, Y. Ma, and A.M. Yucheng Lan, Bo Yu, Xiao Yan, Dezhi Wang, Andrew Muto, Daryoosh Vashaee, Xiaoyuan Chen, Junming Liu, Mildred S. Dresselhaus, Gang Chen, and Zhifeng Ren, Science **320**, 634 (2008).
4. A.F. Ioffe, *Semiconductor Thermoelements and Thermoelectric Cooling*, (Infosearch Ltd., London, UK, 1957).
5. H.J. Goldsmid, *Thermoelectric Refrigeration*, (Plenum, New York, 1964).
6. N. Savvides and H.J. Goldsmid, J. Phys. C **6**, 1701 (1973).
7. D.M. Rowe, V.S. Shukla, and N. Savvides, Nature **290**, 765 (1981).
8. U. Winkler, Helv. Phys. Acta **28**, 633 (1955).
9. R.G. Morris, R.D. Redin, and G.C. Danielson, Phys. Rev. **109**, 1909 (1958).
10. R.D. Redin, R.G. Morris, and G.C. Danielson, Phys. Rev. **109**, 1916 (1958).
11. V.K. Zaitsev, M.I. Fedorov, E.A. Gurieva, I.S. Eremin, P.P. Konstantinov, A.Y. Samunin, and M.V. Vedernikov, Phys. Rev. B. **74**, (2006).
12. M. Riffel and J. Schilz, Scripta Mater. **32**, 1951 (1995).
13. T. Aizawa, R. Song, and A. Yamamoto, Mater. Trans. JIM **46**, 1490 (2005).
14. J. Tani and H. Kido, Physica B **364**, 218 (2005).
15. G.S. Nolas, D. Wang, and M. Beekman, Phys. Rev. B **76**, 235204 (2007).
16. Q. Zhang, J. He, T.J. Zhu, S.N. Zhang, X.B. Zhao, and T.M. Tritt, Appl. Phys. Lett. **93**, 102109 (2008).
17. H.Y. Chen and N. Savvides, J. Elec. Mater., in press (2009).
18. C. Chuang, N. Savvides, and S. Li, J. Elec. Mater., in press (2009).
19. N. Savvides and H.Y. Chen, to be published (2009).

Mater. Res. Soc. Symp. Proc. Vol. 1166 © 2009 Materials Research Society 1166-N06-03

On the Thermoelectric Potential of Inverse Clathrates

Matthias Falmbigl[1], Peter F. Rogl[1], Ernst Bauer[2], Martin Kriegisch[2], Herbert Müller[2], Silke Paschen[2]

[1] Institute of Physical Chemistry, University of Vienna, A-1090 Wien, Austria
[2] Institute of Solid State Physics, Vienna University of Technology, A-1040 Wien, Austria

ABSTRACT

In the context of a general survey on the thermoelectric potential of cationic clathrates, formation, crystal chemistry and physical properties were investigated for novel inverse clathrates deriving from $Sn_{19.3}Cu_{4.7}P_{22}I_8$. Substitution of Cu by Zn and Sn by Ni was attempted to bring down electrical resistivity and lower thermal conductivity. Materials were synthesized by mechanical alloying using a ball mill and hot pressing. Structural investigations for all specimens confirm isotypism with the cubic primitive clathrate type I structure (lattice parameters a = ~1.1 nm and space group type Pm-3n). Studies of transport properties evidence holes as the majority charge carriers. Thermal expansion data, measured in a capacitance dilatometer from 4 to 300 K on $Sn_{19.3}Cu_{1.7}Zn_3P_{19.9}\square_{2.1}I_8$, compare well with literature data available for $Sn_{24}P_{19.6}\square_{2.4}Br_8$ and for an anionic type I clathrate $Ba_8Zn_8Ge_{38}$. From the rather complex crystal structure including split atom sites and lattice defects thermal conductivity in inverse clathrates is generally low. Following Zintl rules rather closely inverse clathrates tend to be semiconductors with attractive Seebeck coefficients. Thus for thermoelectric applications the main activity will have to focus on achieving low electrical resistivity in a compromise with still sufficiently high Seebeck coefficients.

INTRODUCTION

Efficient thermoelectric (TE) power generation and thermoelectric cooling technologies request materials of low electrical resistivity, ρ, and low thermal conductivity, λ, but with a high Seebeck coefficient, S [1]. Thermoelectric application orients itself on the so-called (dimensionless) thermoelectric figure of merit, $ZT=S^2T/(\rho\lambda)$, which should exceed ZT>1 for efficient conversion of heat into electric power in a thermoelectric generator (TEG). "Classical" thermoelectric materials based on bismuth and lead tellurides have recently achieved ZT~1.2-1.4 by selective doping, superlattice technologies or nanostructuring but suffer from low thermal stability ($T_m(Bi_2Te_3)=586$ °C; $ZT_{max}(Bi_2Te_3)$ at 100 °C; $ZT_{max}(PbTe)$ at 350 °C) [2,3]. Design of thermoelectric generators for automotive applications, however, needs thermoelectric materials that can reliably function at elevated temperatures [4]. "Intermetallic clathrates", especially compounds with clathrate type I structure, proved that they may fulfill these requirements exploiting the Phonon Glass-Electron Crystal (PGEC) concept developed by Slack based on (a) a sufficiently good electrical conductivity in combination with (b) weakly bound atoms or molecules, which via their "rattling modes" efficiently scatter phonons and thus decrease thermal conductivity [5]. In this respect the complicated crystal structure of clathrates offers a wide spectrum of tools for tuning the electronic structure and corresponding TE properties by performing (a) various atom substitutions (doping) in the framework, regulating the concentration of vacancies as well as by (b) introducing effects of nano-structuring (nano-

Table I. Overview on inverse type I clathrate compounds, structural and physical property data. □ is a symbol for vacancies.

Compound	Structure Type Spacegroup	Lattice parameter [10^-10 m]	Seebeck coefficient [μV/K] (300K)	Electrical cond. (σ) [S/m] (300K)	Thermal cond. (λ) [W/mK] (300K)	ZT	ref.
Si-containing:							
$Si_{46-x}I_xI_8$	Pm-3n	10.4195(7)					32
$Si_{46-x}P_xTe_8$ x= 11			60	50000			33
12			90	30000			
13			135	6000	3.9	0.45 (900K)	
14			210	1000			
15			220	300			
16	Pm-3n	9.96457(9)	-10	0.158	4.4		
17			250 (500K)	0.0158		0.06 (900K)	
$Si_{46-x}P_xTe_y$ x= y=							15
14.7 7.35	Pm-3	9.9702(3)					
13.6 6.98	Pm-3	9.9794(2)					
13.0 6.88	Pm-3	9.9808(2)					
Ge-containing:							
$Ge_{38}P_8X_8$ X= Cl		10.3514(3)					8
Br	P-43n	10.4074(4)					
I	P-43n	10.5067(6)		1.11-6.25			
$Ge_{30}P_{16}Te_8$	Pm-3n	10.3376(2)	750	3.16	0.86		34
$Ge_{38}As_8X_8$ X= Br		10.5161(5)					8
I	Pm-3n	10.625		100			35
I	P-43n	10.6158(4)		0.02-0.008			8
$Ge_{38}Sb_8X_8$ X= Br		10.7893(8)					8
I		10.8697(6)		0.00002			8
I	Pm-3n	10.8892(2)	-800	0.001	1.15	0.001	36
$Ge_{38}Sb_8I_{7.81}$	Pm-3n						37
$Ge_{38+x}Sb_{8-x}I_{8-x}$ (0<x≤0.86)	10.850(3) for x = 0.86						
$Ge_{38-y-y/4}□_{y/4}Sb_{8+y}I_8$ (0<y≤6.83)	10.946(4) for y = 6.83						
$Ge_{14}Ga_{12}Sb_{20}I_8$	P-43n	11.273(2)					38
$Ge_{40.9}Te_{5.3}□_{0.7}I_8$	Pm-3n	10.815(1)					14
$Ge_{43.33}I_{2.67}I_8$							39
Sn-containing:							
$Sn_{24}P_{19.3}□_{2.7}Cl_yI_{8-y}$ y=							
0	Pm-3n	10.954(1)	80	1500	1.8	0.02 (300K)	40, 11
0.8	Pm-3n	10.933(1)					11
$Sn_{24}P_{19.3}□_{2.7}Br_xI_{8-x}$ x=							
0	Pm-3n	10.954(1)	80	1500	1.8	0.02 (300K)	40, 11
2	Pm-3n			1340	0.5	0.02 (300K)	30, 11
3	Pm-3n			545			30, 11
7	Pm-3n			405			30, 11
8	Pm-3n	10.8142(7)	180	334			30, 11
$Sn_{17}Zn_7P_{22}Br_8$	Pm-3n	10.7254(2)		250			23
$Sn_{19.3}Cu_{4.7}P_{22}I_8$	Pm-3n	10.847(1)					20
$Sn_{20}Zn_4P_{20.8}□_{1.2}I_8$	Pm-3n	10.881(1)					23
$Sn_{17}Zn_7P_{22}I_8$	Pm-3n	10.8425(6)		0.25			23
$Sn_{14}In_{10}P_{21.2}□_{0.8}I_8$	P4 2/m	a=24.745(3); c=11.067(1)					10
$Sn_{10}In_{14}P_{22}I_8$	Pm-3n	11.0450(7)					10
$Sn_{24}As_{19.3}□_{2.7}I_8$		22.179(1)					40
$Sn_{20.3}□_{3.5}As_{22}I_8$	Pm-3n	11.092(1)	-180	0.95	0.5	2*10^-5 (300K)	12
	F23 / Fm-3	22.1837(4)					12
$Sn_{19.3}Cu_{4.7}As_{22}I_8$	Pm-3n	11.1736(3)					19
$Sn_{38}Sb_8I_8$	Pm-3n	12.0447(3)	-600	0.1	0.7		36

precipitates out of temperature dependent solvus surfaces, nano-dispersions of foreign phases in composites, etc.).

Besides the general class of intermetallic clathrates, which consist of cationic guest atoms in the voids of a covalently bonded framework (incorporating anionic elements in order to balance the host charges) [6], there exists a growing group of so-called "inverse clathrates" with reversed polarity in the host framework [7]. Following the Zintl concept for semiconductors, Menke and von Schnering [8] were the first to synthesize polycationic clathrates, $Ge_{38}A_8X_8$, X = Cl, Br, I; A = P, As, Sb. Since then, there have been successful efforts to substitute the group 14 elements (Si, Ge, Sn) in inverse clathrates partially by P, As, Sb, Ga, Te, I, In, Zn, Cd, Cu [9]. In recent years various studies have shown that Sn-based inverse clathrates derive from an ideal composition $Sn_{24}A_{22-y}\square_yX_8$ (A = P, As, Sb; X = I, Br, Cl, Te), which on substitution may develop a varying amount of vacancies in the framework [10,11,12].

Although only some of the aforementioned compounds are stable at elevated temperatures, for many of the hitherto synthesized inverse clathrates the thermoelectric properties (S, λ, ρ) have not been investigated so far as can be seen from the compilation in Table I.

DISCUSSION

On the crystal chemical relations between intermetallic and inverse type I clathrates

The clathrate type I structure consists of a framework made by 24-vertex tetrakaidecahedra [$5^{12}6^2$] which are linked to each other by sharing hexagonal faces forming non-intersecting channels, which enclose isolated 20-vertex pentagondodecahedra [5^{12}]. This type I framework contains 46 atoms E. Filler (host) atoms G are incorporated (clathrated) in 2 small cavities (at the centers of pentagondodecahedra) and 6 larger cavities (at the centers of tetrakaidecahedra) resulting in a general formula $G_8[E_{46}]$ (for details see ref. [6]). Structures of type I generally crystallize in the space group Pm-3n (Pearson symbol cP54). Although for structures $Ge_{38}A_8X_8$ (A = P, As, Sb and X = Cl, Br, I) the centre of symmetry was claimed to be absent and they all were reported to adopt the space group P-43n [8], a recent redetermination of the structure of $Ge_{38}Sb_8I_8$ unambiguously revealed centro-symmetry Pm-3n and thus isotypism with clathrate type I [13]. Although the majority of hitherto known inverse clathrates adopts the type I structure with space group Pm-3n, several structure variants have been encountered which seem to all derive from parent Pm-3n symmetry. Figure 1 summarizes for all hitherto known inverse clathrates the crystallographic group-subgroup relationships among the aristo-type and the lower symmetry structures derived from it. The driving force to superstructures is atom ordering (such as observed for $Ge_{38}A_8X_8$ [8]; $Ge_{40}Te_{5.3}\square_{0.7}I_8$ [14]; $Si_{46-x}P_xTe_y$ [15]; $Sn_{24}A_{22}X_8$ with A = P, As, Sb and X = I, Br, Cl, Te [9]; $Ba_8Cu_{16}P_{30}$ [16]), but also defect ordering plays a significant role such as in $Ba_8Ge_{43}\square_3$ [17] or in $Sn_{14}In_{10}P_{21.2}\square_{0.8}I_8$ [10]. The manifold of superstructures is at present distinctly higher for cationic clathrates than for anionic counterparts where we only can list so far vacancy ordering for $Ba_8Ge_{43}\square_3$, enlarging the unit cell 8-fold ($a=2a_0$) with respect to the parent type.

Most clathrate compounds are semiconductors and thus strictly obey the Zintl formalism ([for details see ref. 18]), which in turn is a simple tool to either design novel compositions for substituted clathrates and/or to calculate compensation by vacancy formation in the framework structure.

Within clathrate compounds a group can be distinguished, which lies beyond the scope of the Zintl formalism. The compounds involved are characterized by their vacancy free structures. The electrons are not compensated and so they occupy the anti-bonding conduction band of the host framework and exhibit metallic properties. This deviation of the Zintl formalism only appears when the cleavage of the bonds of the framework atoms is energetically less favorable than the filling of the conduction band. Also the tin arsenide iodide "$Sn_{17.6}Cu_{6.4}As_{22}I_8$" does not satisfy the Zintl principle, but references for this compound are contradictory. In [9], this compound is presented as an example for deviations of the Zintl formalism, but in [19] it is pointed out, that all attempts synthesizing a sample with the composition $Sn_{24-x}Cu_xAs_{22}I_8$ where x > 4.7 were unsuccessful. The reasons for the deviation stay unclear. Maybe also in this case the vacancy formation is energetically less favorable than filling the anti-bonding states with excess electrons.

Clathrates Type I
Symmetry Variants

Group - Subgroup Relations

Compound	Space gr.
$Ba_8M_xGe_{46-x-y}\square_y$ parent type	$P4/m\bar{3}2/n$
$Ba_8Ge_{43}\square_3$	$I4_1/a\bar{3}2/d$
$I_8Sn_{20.5}As_{22}\square_{3.5}$	$Fm\bar{3}$ or $F23$
$I_8Sb_8Ge_{38}$ isotypic (Cl,Br,I)$_8$(P,As,Sb)$_8Ge_{38}$	$P\bar{4}3n$
$Te_{8-y}Si_{46-x}P_x$	$P2/m\bar{3}$
$I_8Ge_{40}Te_{5.2}\square_{0.8}$	$P23$
$I_8Sn_{14}In_{10}P_{21.2}\square_{0.8}$	$P4_2/m$

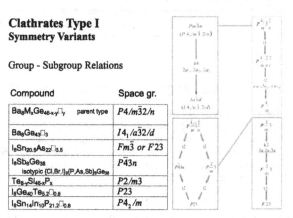

Figure 1. Crystallographic group-subgroup relations among clathrate type I superstructures.

As a consequence of vacancy formation atom positions neighbouring the vacancies usually are described by crystallographic split positions taking care of the fact that the presence of an atom or of a void in a site defines a slightly different bonding distance to the surrounding framework atoms. In most clathrate type I structures the 6d site is the site preferred by vacancies irrespective of anionic or cationic clathrate compounds [6,7]. Accordingly the adjacent 24k site appears in form of a split site where only one of them is occupied at a time vanishing in correlated manner with the disappearance of the void in 6d. Note that we refer to crystallographic standardized atom positions (using program Structure Tidy [22]).

Synthesis and structure of new clathrate I compounds $Sn_{19.3}Cu_{4.7-x}Zn_xP_{22-y}\square_yI_8$ (x = 1, 2, 3), and "$Sn_{20.5}Ni_{3.5}P_{22}I_8$"

In the present work we focused on the compound $Sn_{19.3}Cu_{4.7}P_{22}I_8$ [20] with the aim (a) to find a simple procedure for synthesis, (b) to investigate the thermoelectric properties and (c) to optimize the thermoelectric performance via substitution of Cu and Sn by other suitable elements. All samples were prepared in form of 95% dense cylinders (10 mm diameter, 8 mm

162

height) via ball-milling in a 45 ml steel vessel of a "Fritsch-Pulverisette 4" planetary mill under cyclohexane (250 rpm main disc speed, ratio –2.5, 15 repetitions for 10 min ball milling and 6 min rest in between each milling step) followed by short term hot-pressing under Ar at T_{max}=500 °C (30 min under 4 kN). Starting materials were 99.9 % pure powders of Sn, Ni, Cu, Zn, red P and SnI$_4$. The obtained materials were checked by SEM-EDX and XPD and revealed practically single-phase condition for a type I clathrate with some small amounts of fine-grained grain boundary phases.

Atom substitution of Cu by Zn should lead to a change in the lattice parameters (see figure 2). When Zn substitutes for Cu, the lattice parameter shows an almost linear increase up to the nominal composition Sn$_{19.3}$Cu$_{1.7}$Zn$_3$P$_{22-y}\Box_y$I$_8$. From Rietveld-refinement the Zn-containing samples form vacancies in the 6d site of phosphorus depending on the amount of substitution, as shown in figure 2. These vacancies are somehow overcompensating the additional electrons coming from the Zn.

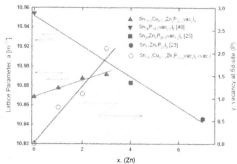

Figure 2. Variation of lattice parameter a with Zn-content in the samples Sn$_{19.3}$Cu$_{4.7-x}$Zn$_x$P$_{22-y}\Box_y$I$_8$ (x = 1,2,3) (filled triangles) and for Sn$_{24-x}$Zn$_x$P$_{22-y}\Box_y$I$_8$, x=0, 4, 7 (filled symbols); dependency of vacancy formation on the Zn-content: open circles.

Thermoelectric performance of cationic clathrates

Following the Zintl rules, both anionic and cationic clathrates are supposed to reveal semiconducting properties, i.e, high electrical resistivity and concomitantly large Seebeck coefficients. Table I essentially confirms this scheme. Starting from the clathrate Sn$_{19.3}$Cu$_{4.7}$P$_{22}$I$_8$, described in [20], attempts were made to increase its electrical conductivity via Cu/Zn substitution, which should introduce with every Zn-atom an additional electron. Therefore a series of samples Sn$_{19.3}$Cu$_{4.7-x}$Zn$_x$P$_{22-y}\Box_y$I$_8$ (x = 1, 2, 3) was investigated. From Rietveld-refinement the vacancies could be located in the phosphorus 6d site, which is in agreement with the so far investigated clathrate I compounds yielding y=0.8 (0.2) for x=1, y=1.1 (0.4) for x=2 and y=2.1 (0.6) for x=3. (In brackets the theoretical number of vacancies in the P 6d site according to Zintl-formalism is given).

In figure 3 the electrical resistivities $\rho(T)$ of all samples are compared to the values of Sn$_{17}$Zn$_7$P$_{22}$I$_8$ [23]. Common to all clathrates investigated is a significant increase of $\rho(T)$ with decreasing temperature as typical for semiconductors with thermally activated conductivity. A standard fit in terms of equation 1

$$\rho(T) = \rho_0 + C \exp\frac{\Delta E_{gap}}{2k_B T} \tag{1}$$

reveals good agreement for band gaps ΔE_{gap} ranging from 0.29 eV for $Sn_{19.3}Cu_{3.7}Zn_1P_{21.2}\square_{0.8}I_8$ to 0.26 eV for $Sn_{19.3}Cu_{2.7}Zn_2P_{20.9}\square_{1.1}I_8$ and to 0.21 eV for $Sn_{19.3}Cu_{1.7}Zn_3P_{19.9}\square_{2.1}I_8$. These values fit quite well to $\Delta E_{gap} = 0.25$ eV reported for $Sn_{17}Zn_7P_{22}I_8$ [23]. In equation 1 ρ_0 is the resistivity at infinitely high temperature and C is a constant.

A closer inspection of the data, however, demonstrates that a description of the resistivity data by Mott's model of hopping conductivity [24] is preferable as typical for conductivity of localized electron systems. A precondition for this hopping mechanism is the overlap of the wave functions of an occupied and an unoccupied state. The temperature dependency of the electrical conductivity is then described by:

$$\sigma = \sigma_0 \exp\left[-\left(\frac{T_0}{T}\right)^{\frac{1}{n}}\right] \quad \text{with} \quad T_0 = \frac{(4\alpha)^3}{9\pi N(E_F)k_B} \tag{2}$$

Here α is a parameter for the decay of the wave function, $N(E_F)$ the density of states at the Fermi level and σ_0 is a constant. The exponent n is usually set to $n = 4$. In case of strong Coulomb correlations, however, n adopts a value of two for all dimensions. The competition between disorder and the Coulomb interaction energy leads to a depletion of the single particle density of states near the Fermi energy, known as Coulomb gap. Fitting various types of hopping clearly reveals a better agreement for $n = 2$. Figure 3 compares the activation type model with the hopping mechanism for $n = 2$.

In figure 4 the temperature dependence of the Seebeck coefficient, S, is summarized for selected inverse clathrates, all showing positive values, characteristic for p-doped semiconductors. Obviously, two different scenarios can be identified: a) extremely large Seebeck coefficients are found for those clathrates exhibiting highest resistivity values.

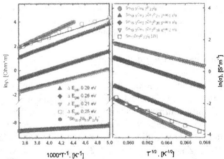

Figure 3. Comparison of electrical resistivity/conductivity of samples $Sn_{19.3}Cu_{4.7-x}Zn_xP_{22-y}\square_yI_8$ in terms of an activation type model (left panel) with the hopping mechanism for n = 2 described in text (right panel).

As typical for semiconductors at high temperatures, S(T) tends to decrease with temperature as a consequence of the thermally populated conduction band; b) clathrate samples behaving more metallic like, show a steady increase of S(T) with rising temperature due to an almost constant charge carrier density. In the latter case, S(T) is proportional to temperature but inversely proportional to the charge carrier density.

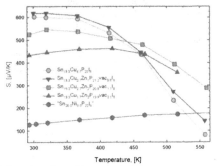

Figure 4. Seebeck coefficient for selected inverse clathrates.

Thermal transport

As there are no rigid bonds between the clathrate framework and the guests, these filler atoms can serve as efficient rattlers reducing the phononic part of the thermal conductivity. At room temperature the heat conductivity of many type I clathrates is $0.5–1.5$ $Wm^{-1}K^{-1}$, which is comparable to that of amorphous compounds. Calculations for clathrate compounds based on the Wiedemann-Franz law demonstrated that the electronic component of the heat conductivity is not higher than 6–8 %. Consequently, the heat conductivity in these materials is dominated by the contribution of the lattice, which is lowered by rattling of loosely bonded guest atoms. The resulting thermal conductivity depends on factors like the guest atom size and mass, the host-guest size matching, and the possibility that guest atoms of different kind can be interchanged in the framework cages. In the general case, the thermal conductivity decreases as the guest atom mass and the volume free for its oscillations increase, and with the increase in the degree of disordering in the distribution of two different guest atoms [7]. Iodine is the heaviest anionic guest, which can be incorporated into a clathrate compound.

The temperature dependent thermal conductivity was measured from 4 K to room temperature for a sample where we attempted to introduce Ni into the framework structure: "$Sn_{20.5}Ni_{3.5}P_{22}I_8$". Based on the Wiedemann-Franz law, the data observed were separated into a lattice (λ_{ph}) and an electronic contribution (λ_e) where $\lambda = \lambda_{ph} + \lambda_e$. With the large electrical resistivity, $\lambda_e(T)$ is rather small, and near room temperature stays well below a few percent of λ_{ph}. No low temperature maximum in $\lambda_{ph}(T)$ is observed, in contrast to Zn, Cd, and Pd-rich clathrates $Ba_8\{Zn,Cd,Pd\}_xGe_{46-x-y}\square_y$ [25]. A quantitative description of $\lambda_{ph}(T)$ was performed applying Callaway's theory [26] of lattice thermal conductivity (equation 3), which follows from the basic thermodynamic expression $\lambda=(1/3)C_V$ v l, where C_V is the heat capacity of the system, v the particle velocity and l the mean free path,

$$\lambda_{ph} = \frac{k_B}{2\pi^2 v_s}\left(\frac{k_B}{\hbar}\right)^3 T^3 \int_0^{\theta_D/T} \frac{\tau_C x^4 e^x}{(e^x-1)^2}dx \qquad (3)$$

with $v_s=(k_B\theta_D)/(h(6p^2n)^{1/3})$ the velocity of sound in terms of the Debye model, n is the number of atoms per unit volume and $x = h\omega/k_BT$ with ω the phonon frequency. $1/\tau_c$ stands for the sum of the reciprocal relaxation times for point defect scattering, Umklapp processes, boundary scattering and scattering of phonons by electrons. τ_c does not contain terms for resonance scattering, since they seem to be of minor importance here. A T^3 term was added to compensate for radiation losses (inherited from the steady state heat flow method used). The initial rise of λ_{ph} stems from a combination of boundary and point defect scattering, including vacancies. The missing maximum in $\lambda_{ph}(T)$ at low temperatures is thus a sign of substantial scattering from both mechanisms.

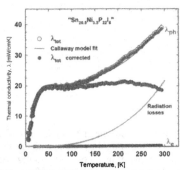

Figure 5. Lattice $\lambda_{ph}(T)$ and electronic $\lambda_e(T)$ thermal conductivity of "$Sn_{20.5}Ni_{3.5}P_{22}I_8$". The solid line is a least squares fit according to equation 3 including a T^3 correction for radiation losses. "λ_{tot}-corrected" shows the total thermal conductivity as measured minus the T^3 correction.

A least squares fit to equation 3, shown as a solid line in figure 5, reveals fine agreement with the experimental data and evaluation of terms indicates that point defect scattering on atom disorder and vacancies is the prominent contribution. An approximation according to Cahill and Pohl [27], allows estimating the theoretical lower limit of lattice thermal conductivity, which primarily is defined by the number of atoms per unit volume and by the Debye temperature. Taking $N = 4.3*10^{28}$ m^{-3} and $\theta_D \sim 200$ K reveals a $\lambda_{min}(300$ K$) \sim 4$ mW/cmK.

Figure 6 presents an overview on the lattice thermal conductivities $\lambda_{ph}(T)$ for clathrates in comparison with various other thermoelectric materials. Cationic clathrates span a large region from high thermal conductivities for "light frameworks" $(Si,P)_{46}Te_8$ to rather low values for disordered lattices with heavier Sn-based frameworks almost reaching values of amorphous materials.

From measurements of ρ, λ and S the thermoelectric figure of merit $ZT=S^2T/(\rho\lambda)$ was extracted for "$Sn_{20.5}Ni_{3.5}P_{22}I_8$" (the material for which electrical resistivity was lowest). The values for ZT are still small although the thermal conductivity is at a good level and the Seebeck coefficient is very high, but a still insufficient electrical conductivity diminishes the figure of merit. Figure 7 reveals thermoelectric figure of merit ZT for "$Sn_{20.5}Ni_{3.5}P_{22}I_8$" in comparison with various thermoelectric materials. Figures 6,7 also document that for cationic clathrates presently more can be gained for the figure of merit by engineering lower electrical resistivities than by lowering thermal conductivity.

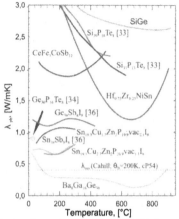

Figure 6. Overview on the lattice thermal conductivity $\lambda_{ph}(T)$ for various thermoelectric materials including type I clathrates (basic data after Snyder [28]).

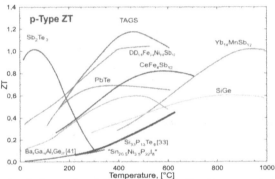

Figure 7. Thermoelectric figure of merit ZT for "$Sn_{20.5}Ni_{3.5}P_{22}I_8$" in comparison with various p-type thermoelectric materials (basic data after Snyder [28]).

Thermal expansion

For practical applications of thermoelectrics, thermal expansion plays a decisive role in engineering of thermal and electric TE contacts. Hitherto very little is known on mechanical properties for inverse clathrates. Figure 8 shows the thermal expansion (measured in a capacitance dilatometer [29]) for $Sn_{19.3}Cu_{1.7}Zn_3P_{19.9}\square_{2.1}I_8$ together with data available for $Sn_{24}P_{19.6}\square_{2.4}Br_8$ [30] and in comparison with thermal expansion for an anionic clathrate $Ba_8Zn_8Ge_{38}$ [25]. It is interesting to note that the thermal expansion coefficients $\alpha_{RT}=\partial(\Delta L/L_0)/\partial T$ (at room temperature RT) are rather similar for all three compounds ranging at about $1.1*10^{-5}$ K^{-1}. Following a semiclassical treatment to calculate the thermal expansion taking into account three and four-phonon interactions considering an anharmonic potential and using the Debye model for the acoustic phonons and the Einstein approximation for the optical modes [31], we obtain from a fit to the equations 4 below the length change $\Delta L/L(T_0)$ [31]:

$$\frac{\Delta L}{L(T_0)} = \frac{\langle x \rangle_T - \langle x \rangle_{T_0}}{x_0} \qquad \langle x_T \rangle = \frac{\gamma}{2} T^2 + \frac{3 g}{4 c^2} [\varepsilon - G \varepsilon^2 - F \varepsilon^3]$$

$$\varepsilon = \left\{ \left(\frac{3}{p} \right) 3 k_B T \left(\frac{T}{\theta_D} \right)^3 \theta_D \int_0^{/T} \frac{z^3 dz}{e^z - 1} + \left(\frac{p-3}{p} \right) \frac{k_B \theta_E}{e^{\theta_E / T} - 1} \right\} \qquad (4)$$

γ is the electronic contribution to the average lattice displacement, θ_D is the Debye temperature, θ_E is the Einstein temperature and p is the average number of phonon branches actually excited over the temperature range. G, F, c, g and f are further material dependent constants.

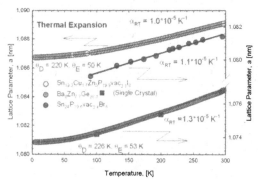

Figure 8. Thermal expansion of $Sn_{19.3}Cu_{1.7}Zn_3P_{19.9}\square_{2.1}I_8$ compared to other clathrate compounds.

As seen from figure 8 reliable fits can be obtained yielding $\theta_D = 226$ K and $\theta_E = 53$ K for $Ba_8Zn_8Ge_{38}$, in excellent agreement with specific heat data [25]. Similarly, a fit to the thermal expansion curve for $Sn_{19.3}Cu_{1.7}Zn_3P_{19.9}\square_{2.1}I_8$ yields a perfect correspondence for $\theta_D = 220$ K and $\theta_E = 50$ K demonstrating the close behavior of anionic and cationic clathrate compounds of type I. The low lying optical mode is an archetypical property of cage forming compounds with loosely bound filler elements, occupying the oversized cages.

General remarks on the thermoelectric potential of inverse clathrates

Although several type I compounds among inverse clathrates exhibit remarkably high thermal stability, such as the solution $Si_{46-x}P_xTe_8$ [33], most of the inverse clathrate materials show thermal decomposition already above 600 °C with high vapor partial pressures of iodine and/or phosphorous even at lower temperatures. The rather complex crystal structure of clathrates, where in several cases framework atoms randomly share atom sites with vacancies thereby inferring split atom sites in neighboring coordination shells, results in generally low thermal conductivities for inverse clathrates particularly for framework structures with heavier atoms (such as Sn in contrast to P and/or Si). Obeying Zintl rules rather closely, inverse clathrates appear to be semiconductors with high Seebeck coefficients. As a consequence electrical resistivities can be rather large. Thus for thermoelectric applications the main activity will have to focus on achieving low electrical resistivity either via proper chemical substitutions or via suitable grain boundary phases in a compromise with still sufficiently high bulk Seebeck coefficients.

ACKNOWLEDGMENTS

Financial support from the FFG project THECLA (815648) is gratefully acknowledged.

REFERENCES

1. D. M. Rowe, Ed., *CRC Handbook of Thermoelectrics* (CRC Press, Boca Raton, FL, 2006).
2. R. Venkatasubramanian, E. Siivola, B. O'Quinn, *"Superlattice Thin-Film Thermoelectric Material and Device Technologies"*, CRC Handbook of Thermoelectrics, ed D.M. Rowe (CRC Press, Boca Raton, FL, 2006) pp. 49.1-49.15.
3. Y. Gelbstein, Z. Dashevsky, M.P. Dariel, *Phys. Stat. Sol. (RRL)* **1**(6), 232– 234 (2007).
4. P. Prenninger, A. Grytsiv, P. Rogl, E. Bauer, „*TE-Materials with Better Efficiencies and Lower Costs – a Contradiction?*, paper presented at the 1st Conference: Thermoelectrics – A Chance for the Automotive Industry?, Berlin, 23 – 24 October 2008.
5. G.S. Nolas, J.L. Cohn, G.A.. Slack, S.B. Schujman, *Appl. Phys. Lett.* **73**, 178 (1998)
6. P. Rogl, „*Formation and Crystal Chemistry of Clathrates"*, CRC Handbook of Thermoelectrics, ed D.M. Rowe (CRC Press, Boca Raton, FL, 2006) pp. 32-1 – 32-24.
7. A.V. Shevelkov, *Russian Chemical Reviews* **77**(1), 1-19 (2008).
8. H. Menke, H.G. von Schnering, *Zeitschrift fuer Anorganische und Allgemeine Chemie* **395**(2-3), 223-38 (1973).
9. K.A. Kovnir, A.V. Shevelkov, *Russian Chemical Reviews* **73**(9), 923-938 (2004).
10. M.M. Shatruk, K.A. Kovnir, M. Lindsjoe, I.A. Presniakov, L.A. Kloo, A.V. Shevelkov, *J. Solid State Chem.* **161**, 233-242 (2001).
11. J.V. Zaikina, W. Schnelle, K.A. Kovnir, A.V. Olenev, Y. Grin, A.V. Shevelkov, *Solid State Sciences* **9**(8), 664-671 (2007).
12. J.V. Zaikina, K.A Kovnir, A.V. Sobolev, I.A. Presniakov, Y. Prots, M. Baitinger, W. Schnelle, A.V. Olenev, O.I. Lebedev, G. Van Tendeloo, Y. Grin, A.V. Shevelkov, *Chemistry-A European Journal* **13**(18), 5090-9 (2007).
13. Ya. Mudryk, P. Rogl, C. Paul, S. Berger, E. Bauer, G. Hilscher, C. Godart, H. Noel, *J. Phys. Condens. Matter* **14**, 7991-8004 (2002).
14. K.A. Kovnir, N.S. Abramchuk, J.V. Zaikina, M. Baitinger, U. Burkhardt, W. Schnelle, A.V. Olenev, O.I. Lebedev, G. Van Tendeloo, E.V. Dikarev, A.V. Shevelkov, *Z. Kristallographie* **221**, 527-532 (2006).
15. J.V. Zaikina, K.A. Kovnir, U. Schwarz, H. Borrmann, A.V. Shevelkov, *Z. Kristallographie - New Crystal Structures* **222**(3), 177-179 (2007).
16. J. Duenner, A. Mewis, *Z. Anorg. Allg. Chemie* **621**, 191 (1995).
17. W. Carrillo-Cabrera, S. Budnyk, Y. Prots, Y. Grin, *Z. Anorg. Allg. Chem.* **630**, 7226 (2004).
18. S.M. Kauzlarich, Ed. „*Chemistry, Structure and Bonding of Zintl Phases and Ions"*, Wiley-VCH, N.Y. (1996).

19.   K.A. Kovnir, A.V. Sobolev, I.A. Presniakov, O.I. Lebedev, G. Van Tendeloo, W. Schnelle, Y. Grin, A.V. Shevelkov, *Inorganic Chemistry* **44**(24), 8786-8793 (2005).

20.   M.M. Shatruk, K.A. Kovnir, A.V. Shevelkov, B.A. Popovkin, *Zhurnal Neorganicheskoi Khimii* **45**(2), 203-209 (2000).

21.   N. Melnychenko-Koblyuk, A. Grytsiv, St. Berger, H. Kaldarar, H. Michor, F. Röhrbacher, E. Royanian, E. Bauer, P. Rogl, H. Schmid, G.Giester, *J. Phys. Cond. Mat.* **19**, 046203-26, (2007).

22.   E. Parthé, L. Gelato, B. Chabot, M. Penzo, K. Cenzual, R. Gladyshevskii, *TYPIX standardized data and crystal chemical characterization of inorganic structure types* (Berlin: Springer) (1994).

23.   K.A. Kovnir, M.M. Shatruk, L.N. Reshetova, I.A. Presniakov, E.V. Dikarev, M. Baitinger, F. Haarmann, W. Schnelle, M. Baenitz, Y. Grin, A.V. Shevelkov, *Solid State Sciences* **7**(8), 957-968 (2005).

24.   N.F. Mott, Phil Mag. B19, 835(1984); *The Physics of Hydrogenated Amorphous Silicon Vol. II.* ed. by J. D. Joannopoulos and G. Luckowsky, *Topics in Applied Physics,* **56**(Springer, Berlin Heidelberg 1984), p. 169.

25.   N. Melnychenko-Koblyuk, A. Grytsiv, L. Fornasari, H. Kaldarar, H. Michor, F. Röhrbacher, M. Koza, E. Royanian, E. Bauer, P. Rogl, M. Rotter, H. Schmid, F. Marabelli, A. Devishvili, M. Doerr, G.Giester, *J. Phys. Cond. Mat.* **19**, 216223 1-26 (2007).

26.   J. Callaway, H. C. von Baeyer, *Phys. Rev.* **120**, 1149 (1960).

27.   D. Cahill, R. Pohl, *Solid State Commun.* **70,** 927 (1989).

28.   G.J. Snyder, E.S. Toberer, *Nature Materials* **7**, 105-114 (2008).

29.   M. Rotter, H. Müller, E. Gratz, M. Doerr, M. Loewenhaupt, *Rev. Sci. Instruments* **69**(7), 2742-45 (1998).

30.   K.A. Kovnir, J.V. Zaikina, L.N. Reshetova, A.V. Olenev, E.V. Dikarev, A.V. Shevelkov, *Inorganic Chemistry* **43**(10), 3230-3236 (2004).

31.   G.D. Mukherjee, C. Bansal, A. Chatterjee, *Phys. Rev. Lett.* **76**(11), 1876-1879 (1996).

32.   E. Reny, S. Yamanaka, Ch. Cros, M. Pouchard, *AIP Conference Proceedings* **590**(Nanonetwork Materials), 499-502 (2001).

33.   K. Kishimoto, T. Koyanagi, K. Akai, M. Matsuura, *Japanese Journal of Applied Physics Part 2: Letters & Express Letters* **46**(29-32), L746-L748 (2007).

34.   K. Kishimoto, K. Akai, N. Muraoka, T. Koyanagi, M. Matsuura, *Applied Physics Letters* **89**(17), 172106/1-172106/3 (2006).

35.   T.L. Chu, S.S. Chu, R.L. Ray, *Journal of Applied Physics* **53**(10), 7102-3 (1982).

36.   K. Kishimoto, S. Arimura, T. Koyanagi, *Applied Physics Letters* **88**(22), 222115/1-222115/3 (2006).

37.   Z. Jin, Z. Tang, A. Litvinchuk, A.M. Guloy, *Abstracts of Papers, 235th ACS National Meeting*, New Orleans, LA, United States, April 6-10, (2008).

38.   H. Menke, H.G.von Schnering, *Zeitschrift Anorg. Allg. Chemie* **424**, 108 (1976).

39.   R. Nesper, J. Curda, H.G. von Schnering, *Angew. Chemie* **25**, 369 (1986).

40.   M.M. Shatruk, K.A. Kovnir, A.V. Shevelkov, I.A. Presniakov, B.A. Popovkin, *Inorganic Chemistry* **38**(15), 3455-3457 (1999).

41.   S. Deng, X. Tang, P. Li, Q. Zhang, *Journal of Applied Physics* **103,** 073503(2008).

Mater. Res. Soc. Symp. Proc. Vol. 1166 © 2009 Materials Research Society   1166-N06-07

Ternary Copper-Based Diamond-Like Semiconductors for Thermoelectric Applications

Donald T. Morelli and Eric J. Skoug

Department of Chemical Engineering & Materials Science

Michigan State University

East Lansing, MI 48824

ABSTRACT

Thermoelectric materials can provide sources of clean energy and increase the efficiency of existing processes. Solar energy, waste heat recovery, and climate control are examples of applications that could benefit from the direct conversion between thermal and electrical energy provided by a thermoelectric device. The widespread use of thermoelectric devices has been prevented by their lack of efficiency, and thus the search for high-efficiency thermoelectric materials is ongoing. Here we describe our initial efforts studying copper-containing ternary compounds for use as high-efficiency thermoelectric materials that could provide low-cost alternatives to their silver-containing counterparts. The compounds of interest are semiconductors that crystallize in structures that are variants of binary zincblende structure compounds. Two examples are the compounds Cu_2SnSe_3 and Cu_3SbSe_4, for which we present here preliminary thermoelectric characterization data.

INTRODUCTION

Goodman and Douglas investigated the possibility of using substitution to formulate ternary compounds having structures based on binary compounds [1]. They noted that the ternary compound $CuFeS_2$ maintained the same electron-to-atom ratio as the binary zincblende compounds $CuBr$ and ZnS and thus had a structure that closely resembled the binary zincblende structure, known as the chalcopyrite structure. They also noted that Fe could be replaced by a group III element, as the function of the Fe is to donate three electrons. Later studies by Hahn et al. [2] and Zhuze et al. [3] confirmed that compounds of composition $A^I B^{III} X_2^{VI}$ crystallize in the chalcopyrite structure, which they noted was analogous to the binary zincblende structure but with the crystallographic "c" axis doubled**. Schematically, the chalcopyrite compounds are "built up" from the binary $A^{II}B^{VI}$ compounds by considering two unit cells of $A^{II}B^{VI}$ ($A_2^{II}B_2^{VI}$), and replacing the divalent group II atoms with a univalent group I atom and a trivalent group III atom. This simplified view explains the configuration of the chalcopyrite unit cell as essentially two zincblende unit cells stacked on top of each other.

** A, B, and X represent elements, superscripts denote the groups of the periodic table from which the elements come, and subscripts denote the stoichiometric proportions of each element. This notation will be used throughout the discussion.

Considering three unit cells of $A^{II}B^{VI}$ and substituting group I and group IV elements on the group II cation sites yields a series of compounds with composition $A_2^I B^{IV} X_3^{VI}$. Averkieva et al. reported on the structure of such compounds, where A^I was copper, B^{IV} was germanium or tin, and X^{VI} was selenium or tellurium [4]. While no silver-containing ternary compounds with the zincblende structure were found, several copper-containing ternary compounds (namely Cu_2SnS_3, Cu_2SnSe_3, and Cu_2SnTe_3) were found to crystallize in the sphalerite structure (equivalent to the zincblende structure with two distinct atoms on the cation site). The stoichiometric germanium-containing compounds of this series all crystallize in a tetragonal structure, but for Cu_2GeSe_3 a cubic lattice was assumed in the presence of excess germanium.

Following the same line of reasoning, $A_3^I B^V X_4^{VI}$ compounds can be "built up" by quadrupling the $A^{II}B^{VI}$ unit cell and substituting group I and group V elements on the cation sites, e.g. Cu_3SbSe_4.

Here we present results on some preliminary investigations of the thermoelectric properties of several copper-containing ternary semiconductors. These compounds may be attractive alternatives to thermoelectric materials like $AgSbTe_2$ and $PbTe$ because they contain neither high-cost silver or toxic lead.

EXPERIMENT

Samples of Cu_2SnSe_3, Cu_3SbSe_4, and Cu_3BiSe_3 were fabricated using standard vacuum melting techniques. Stoichiometric quantities of the elements were placed in a quartz tube and sealed under high vacuum (10^{-6} Torr). Because of the high vapor pressure of Se the ampoules were heated slowly (0.3 K min^{-1}) to 900 C, left for 12 hours, and quenched to room temperature. Resulting ingots were found to be single phase in agreement with previous structural studies of these compounds.

Samples for thermal conductivity, electrical resistivity, and Seebeck coefficient were prepared by cutting parellipipeds of nominal dimensions 4x4x8mm^3 from the as grown boules. Electrical resistivity measurements were performed using a four-probe DC technique and thermal conductivity and Seebeck measurements were made using a standard steady state technique. We estimate an uncertainty in the electrical resistivity of 5 % and an uncertainty in the Seebeck coefficient and thermal conductivity of 10 %.

RESULTS AND DISCUSSION

Cu₂SnSe₃

Our X-ray diffraction studies of this compound (not shown) indicate that it crystallizes in the cubic sphalerite structure with lattice constant of 0.569 nm. No secondary phases were detected in the pattern. Figure 1 shows the resistivity and Seebeck coefficient of our sample of Cu_2SnSe_3.

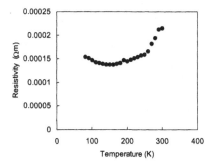

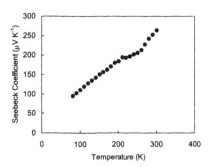

Figure 1. Electrical resistivity and Seebeck coefficient of Cu₂SnSe₃.

The magnitude of resistivity and Seebeck coefficient for this sample identifies it as a semiconductor. Figure 2 shows the thermal conductivity of our sample of Cu₂SnSe₃. Because of the magnitude of the electrical resistivity, thermal conductivity is dominated by the lattice contribution. The behavior is typical of that of a crystalline solid and unlike the thermal conductivity of AgSbTe₂; the significance of this will be discussed below.

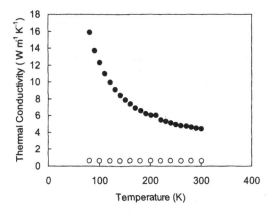

Figure 2. Thermal conductivity of Cu₂SnSe₃. The thermal conductivity of AgSbTe₂ (open symbols) is shown for comparison.

Cu₃SbSe₄

Figure 3 shows the resistivity and Seebeck coefficient of our several samples of Cu₃SbSe₄-based compounds. Three of these samples are nominally undoped, while the fourth was doped p-type with 10 % Sn on the Sb site.

Again we discern from the magnitude of the resistivity and Seebeck coefficient that these

173

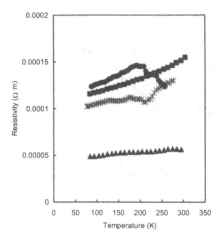
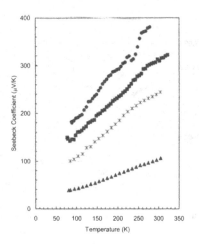

Figure 3. Electrical resistivity and Seebeck coefficient of Cu_3SbSe_4. The three highest curves are for nominally undoped specimens; the lowest curve (filled triangles) is of composition $Cu_3Sb_{0.9}Sn_{0.1}Se_4$.

compounds are semiconducting. The variability in the results for the undoped samples is likely due to self-doping by intrinsic defects. Substitution of Sn for Sb lowers the resistivity and Seebeck coefficient considerably, indicating that this sample is doped more heavily p-type.

Figure 4 shows the thermal conductivity of these fours samples, also compared to $AgSbTe_2$. Again we see typical crystalline behavior in the thermal conductivity, unlike that found for the I-V-VI$_2$ compound.

The results on thermal conductivity of $AgSbTe_2$ suggest that this compound displays "minimal" thermal conductivity. This has been attributed to the nature of bonding in this compound [5]. In the chalcopyrite $A^IB^{III}X_2^{VI}$ compounds, the coordination is tetrahedral and all the outer-shell electrons participate in sp^3 hybridized bonds; their interaction is weak and phonon propagation is largely unaffected. Conversely, in $AgSbTe_2$ the coordination is octahedral, not all of the external s and p electrons in

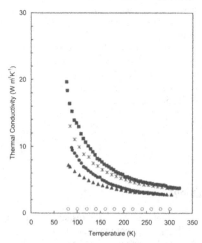

Figure 4. Thermal conductivity of Cu_3SbSe_4; sample designation as in Figure 3.

$A^IB^VX_2^{VI}$ compounds participate in bonding, and the Coulomb interaction between neighboring ions is much stronger, thus disrupting phonon propagation. This is manifested by a large Grueneisen parameter in these compounds [5, 6]. Similarly, in the $A^I_2B^VX^{VI}_3$ and $A^I_3B^VX^{VI}_4$ compounds all of the outer shell electrons participate in bonding and the thermal conductivity is high [7].

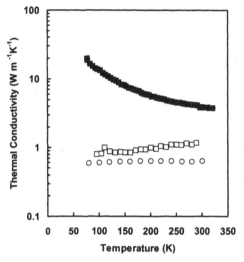

Central to this argument is the formal valence of the group V atom. In the $A^IB^VX_2^{VI}$ compounds the formal valence is 3^+ whereas in $A^I_2B^VX^{VI}_3$ and $A^I_3B^VX^{VI}_4$ compounds it is 5^+. An interesting test of this hypothesis is provided by the compound Cu_3BiSe_3. Here we expect the group V to possess the formal valence of 3^+, and we can compare the thermal conductivity of this compound with Cu_3SbSe_4, where the formal valence of Sb is 5^+. The results are shown in Figure 5. Indeed we find that for the former compound, with a nominally trivalent group V atom, the thermal conductivity is low and similar to $AgSbTe_2$. The trivalent state of Sb leads to "extra" electrons in the hybridized sp^3 bonds in this compound. These unbonded electrons cause electron-cloud repulsion between neighboring ions, leading to large Grueneisen parameter and low thermal conductivity.

Figure 5. Thermal conductivity of Cu_3SbSe_4 (closed squares), Cu_3BiSe_3 (open squares), and $AgSbTe_2$ (open circles).

CONCLUSIONS

We have measured thermal and electronic properties on some ternary Cu-based semiconducting compounds. These compounds offer alternatives to $AgSbTe_2$ and PbTe-base thermoelectric materials but have been little-studied. We find semiconducting behavior and high Seebeck coefficients for both Cu_2SnSe_3 and Cu_3SbSe_4. Dimensionless figure of merit ZT for these samples is in the range 0.02-0.1 at room temperature, and it is expected that higher values of ZT can be obtained at high temperature and with optimization of the doping level. We show that in these ternary systems the magnitude of the thermal conductivity depends on the formal valence of the Group V atom; trivalent Sb or Bi has two unbound electrons in the hybridized bonds that lead to large neighboring-ion Coulomb repulsion effects, large Grueneisen parameter, and resultant low thermal conductivity.

ACKNOWLEDGMENTS

We acknowledge a grant from the National Science Foundation (NSF- CBET-0754029) for partial support of this work.

REFERENCES

1. C.H.L. Goodman and R.W. Douglas, Physica (Amsterdam) **20**, 1107 (1954)
2. H. Hahn, G. Frank, W. Klinger, A.D. Meyer, G. Storger, Z. Anorg. Allg. Chem. **271**, 153 (1954)
3. V.P. Zhuze, V.M. Sergeeva, E.L. Shtrum, Soviet Physics Technical Physics **3**, 10 (1958)
4. G.K. Averkieva, A.A. Vaipolin, N.A. Goryunova, Soviet Research in New Semiconductor Materials, (1965)
5. D.T. Morelli, V. Jovovic, J.P. Heremans, Physical Review Letters **101**, 035901 (2008)
6. D.T. Morelli and G.A. Slack, in High Thermal Conductivity Materials, ed. S.L Shinde and J.S. Goela (Springer, New York, 2006), p. 37.
7. A.V. Petrov, E.L. Shtrum, Soviet Physics – Solid State **4**, 6 (1962)

Mater. Res. Soc. Symp. Proc. Vol. 1166 © 2009 Materials Research Society   1166-N06-09

The Effect on Thermoelectric Properties of Cd Substitution in PbTe

Kyunghan Ahn[1], Mi-Kyung Han[1], Derek Vermeulen[2], Steven Moses[2], Ctirad Uher[2], and Mercouri G. Kanatzidis[1]

[1]Department of Chemistry, Northwestern University, Evanston, IL 60208, U.S.A.
[2]Department of Physics, University of Michigan, Ann Arbor, MI 48109, U.S.A.

ABSTRACT

A recent theoretical study suggested that the substitution of Cd in PbTe can result in a distortion in the electronic density of states (DOS) near the bottom of the conduction band in PbTe. In this study we explored the effect of Cd doping on the thermoelectric properties of PbTe in an effort to test the theoretical prediction that DOS distortion can increase the Seebeck coefficient. We present detailed investigation of structural and spectroscopic data, transmission electron microscopy, as well as transport properties of samples of PbI_2 doped PbTe-x% CdTe (x = 1, 3, 5, 7, 10). All samples follow the Pisarenko relationship and no enhancement of the Seebeck coefficient was observed due to DOS distortions. A low lattice thermal conductivity was achieved by nanostructuring observed via high resolution transmission electron microscopy. A maximum ZT of ~1.2 at ~720 K was achieved for the 1% CdTe sample.

INTRODUCTION

Direct thermal-to-electric energy conversion using thermoelectric materials has been highlighted for the potential of fuel-efficiency improvement in passenger cars and trucks as well as power generation from waste heat generated by industrial plants.[1],[2] The dimensionless thermoelectric figure of merit, $ZT = (\sigma S^2 T)/\kappa$, defines the efficiency of a thermoelectric material, where σ is the electrical conductivity, S is the Seebeck coefficient, T is the absolute temperature, and κ is the thermal conductivity. Significant enhancements in ZT have been recently achieved for both bulk materials and low-dimensional materials.[3],[4],[5]

A recent theoretical study showed that resonance levels can be achieved when Cd is introduced on the metal sites of the PbTe rock salt structure [6] and this can lead to a significant enhancement in the electronic density of states (DOS). Most recently, Heremans et al. [7] reported that values of ZT for p-type $Tl_{0.02}Pb_{0.98}Te$ reach 1.5 at 773 K due to an enhancement in the Seebeck coefficient through a distortion of DOS in the valence band. The substitution of Cd in PbTe is predicted to result in a distortion in the DOS near the bottom of the conduction band and thus can cause an enhancement in the Seebeck coefficient. In addition to resonance levels, a low lattice thermal conductivity can be realized through nanostructuring by the precipitation of CdTe due to a maximum solid solubility limit of CdTe in PbTe according to the PbTe-CdTe pseudobinary phase diagram[8].

In this study we explored the effect of Cd substitution on the thermoelectric properties of PbTe in an effort to test these theoretical predictions. We present detailed investigations of structural properties using spectroscopic measurements and transmission electron microscopy studies as well as transport property investigations including electrical conductivity, Seebeck coefficient, Hall effect, and thermal conductivity on samples of PbTe-0.055% PbI_2-x% CdTe (x = 1, 3, 5, 7, and 10). An enhancement in ZT is observed and its origin is discussed.

EXPERIMENT

Polycrystalline 0.055 mol % PbI$_2$-doped PbTe-x% CdTe (x = 1, 3, 5, 7, 10) ingots of ~10 g were prepared by heating the evacuated quartz tubes encasing graphite crucibles containing PbI$_2$-doped PbTe and CdTe. These tubes were heated at 1323 K for 5 h, and rapidly cooled to 1023 K in 2 h. After dwelling at 1023 K for 4 h, the tubes were cooled to room temperature over 12 h. Dense ingots were obtained with a dark silvery metallic shine. The ingots are stable in water and air and are relatively strong.

Powder X-ray diffraction patterns for samples of PbTe-0.055% PbI$_2$-x% CdTe were collected using a Cu Kα radiation on an INEL diffractometer equipped with a position sensitive detector and operating at 40 kV and 20 A. Room temperature optical diffuse reflectance measurements were performed using a Nicolet 6700 FTIR spectrometer in order to probe optical energy band gaps.

The electrical conductivity σ and Seebeck coefficient S were measured simultaneously from room temperature to ~720 K using a ULVAC-RIKO ZEM-3. The thermal conductivity κ was determined from room temperature to ~720 K through the flash diffusivity method using a NETZSCH LFA 457. The Hall coefficient was measured by using a home-made high temperature apparatus, which provides a working range from 300 K to ~873 K.

Transmission electron microscopy was performed on samples prepared by conventional methods using a JEOL 2100F electron microscope.

DISCUSSION

The powder X-ray diffraction patterns of samples of PbTe-0.055% PbI$_2$-x% CdTe (x = 1, 3, 5, 7, 10) in Figure 1(a) show that these crystallize in the NaCl-type structure without any noticeable second phase. The metallic radius of Pb (1.750 Å) is larger than that of Cd (1.568 Å) and thus a small lattice contraction is expected with Cd incorporation. The refined lattice parameters for x = 1, 3, 5, 7, and 10 are 6.466(1), 6.454(1), 6.451(1), 6.455(1), 6.454(1) Å, respectively. This indicates that there is a maximum solid solubility limit of ~3 % CdTe in PbTe because of nearly constant lattice parameters above x = 3. The diffraction peaks from CdTe (ZnS-type) should be observed for x = 5, 7, 10, but they were not seen because the first and second highest peaks of the ZnS-type phase are overlapped with those of the NaCl-phase. Typical infrared absorption spectra for samples of PbTe-0.055% PbI$_2$-x% CdTe (x = 1, 3, 5, 7, 10) are shown in Figure 1(b). All samples exhibit spectroscopically observable energy band gaps between 0.31 and 0.38 eV compared to ~0.27 eV of pure PbTe.

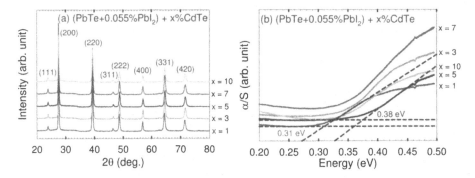

Figure 1. (a) The powder X-ray diffraction patterns (Cu Kα radiation), and (b) Infrared absorption spectra and energy band gaps of samples of PbTe-0.055% PbI$_2$-x% CdTe (x = 1, 3, 5, 7, 10).

Figure 2(a) shows the temperature dependence of the electrical conductivity of samples of PbTe-0.055% PbI$_2$-x% CdTe (x = 1, 3, 5, 7, 10). For all samples, the electrical conductivity decreases with increasing temperature, which is indicative of degenerate doping. The room temperature electrical conductivities of samples with x = 1, 3, 5, 7, and 10 are ~2900, ~1700, ~1500, ~1200, and ~1900 S/cm, respectively. It is evident that the electrical conductivity decreases with increasing CdTe concentration until x = 7 but increases at x = 10. The electrical conductivity of samples follows a temperature dependent power law of $\sigma \approx T^{-\delta}$ ($\delta = 2.1 - 2.3$) which is typical for doped PbTe samples where acoustic phonon scattering dominates at high temperature: δ are 2.3, 2.2, 2.1, 2.1, and 2.3 for x = 1, 3, 5, 7, and 10, respectively.

The Seebeck coefficient of the samples is shown in Figure 2(b). All samples show negative Seebeck coefficients over the entire temperature range indicating n-type conduction. The Seebeck coefficient follows a nearly linear temperature dependence and the room temperature Seebeck coefficients are around -90, -100, -110, -110, and -120 μV/K for x = 1, 3, 5, 7, and 10, respectively. Figure 2(c) shows the temperature dependent power factors ($PF = \sigma S^2$) of samples. As the temperature increases, the power factor increases, reaches a maximum, and then decreases. For instance, in the case of x = 1, the power factor starts from ~22 μW/(cm K^2) at room temperature, reaches a maximum of ~25 μW/(cm K^2) around 430 K, and then decreases to ~17 μW/(cm K^2) around 720 K. The power factors of ~17, ~13, ~13, ~11, and ~13 μW/(cm K^2) were achieved around 720 K for x = 1, 3, 5, 7, and 10, respectively.

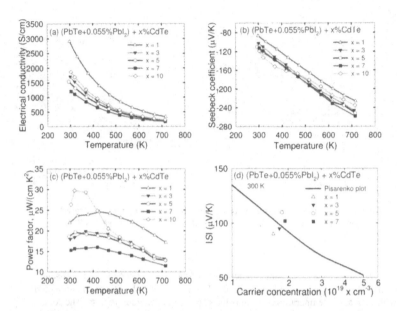

Figure 2. (a) The Electrical conductivity as a function of temperature, (b) The Seebeck coefficient as a function of temperature, (c) The power factor as a function of temperature, and (d) The absolute value of Seebeck coefficient as a function of carrier concentration at room temperature for samples of PbTe-0.055% PbI$_2$-x% CdTe (x = 1, 3, 5, 7, 10). The solid line in (d) is the Pisarenko plot.[9]

Figure 2(d) shows the absolute value of Seebeck coefficient of samples as a function of carrier concentration at room temperature. The solid line in Figure 2(d) is a Pisarenko plot for bulk PbTe[9] and most of n-type and p-type bulk PbTe samples falls on the line. Our samples fall near this line and thus an enhancement in Seebeck coefficient due to resonance levels was not observed.

The temperature dependences of the total thermal conductivity and lattice thermal conductivity of samples are plotted in Figure 3(a). For x = 1, the total thermal conductivity is ~3.0 W/(m K) at room temperature and it decreases with increasing temperature, reaching ~1.1 W/(m K) at ~720 K. The total thermal conductivity is the sum of the electronic thermal conductivity κ_{elec} and the lattice thermal conductivity κ_{latt}. The electronic thermal conductivity can be calculated from the Wiedemann-Franz law, $\kappa_{elec} = L_0 \sigma T$, where the Lorenz number $L_0 = 2.45 \times 10^{-8}$ V^2/K^2 was taken at its full degenerate value. Subtracting the electronic term from the total thermal conductivity one obtains the lattice thermal conductivity of the samples. For all samples, the derived room temperature lattice thermal conductivities are in the range of 0.5 - 1.3 W/(m K). This is to be compared to the lattice thermal conductivity of ~2.2 W/(m K) for PbTe. Namely, the room temperature κ_{latt} was ~0.9, ~1.3, ~1.3, ~1.0, and ~0.5 W/(m K) for x = 1, 3, 5, 7, and 10, respectively. Figure 3(b) shows the relationship between the carrier mobility and the lattice thermal conductivity at room temperature. It indicates that the carrier mobility of samples decreases with CdTe concentration because certain fractions of CdTe exist as nanocrystals in the PbTe matrix and thus play as both electron and phonon scattering centers. However, it is notable that the lowest lattice thermal conductivity was observed for the 1% CdTe sample. Figure 3(c) shows the thermoelectric figure of merit ZT as a function of temperature. The ZTs increase with increasing temperature. The highest ZT of ~1.2, ~0.9, ~0.9, ~1.0, and ~1.0 at ~720 K were achieved for x = 1, 3, 5, 7, and 10, respectively.

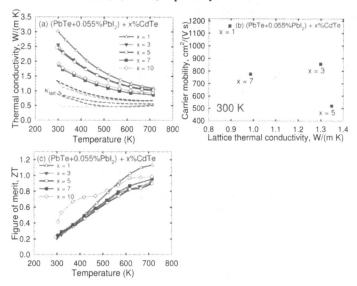

Figure 3. (a) The temperature dependent total thermal conductivity (solid lines) and lattice thermal conductivity (dashed lines), (b) The relationship between the carrier mobility and the lattice thermal conductivity at 300 K and (c) The thermoelectric figure of merit ZTs of samples of PbTe-0.055% PbI_2-x% CdTe (x = 1, 3, 5, 7, 10).

The significant reduction of lattice thermal conductivity may be due to the formation of nanoparticles inside PbTe matrix. It is well known that the presence of coherently embedded nanostructures in the bulk matrix is directly related to the enhanced thermoelectric performance, through reduced lattice thermal conductivity.[3],[10] Low magnification TEM images of the 5% CdTe sample show evenly dispersed nanoparticles in Figure 4(a). As shown in Figure 4(b), the HRTEM images of the 1% CdTe sample indicates a crystalline structure with embedded nanoparticles. The short-range periodicity of the 1% CdTe sample has been evaluated using fast Fourier transform (FFT) analysis, shown in Figure 4(c) and (d). This shows that a typical cubic pattern of the dominant phases with islands with doubling of unit cells. The presence of such nanocrystals inside the sample may be the key for the strong suppression of the lattice thermal conductivity observed in PbTe-x% CdTe in comparison to pure PbTe.

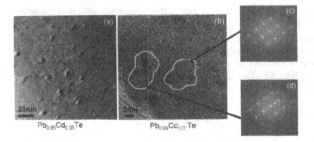

Figure 4. (a) Low magnification TEM of 5% CdTe sample showing various size nanoparticles embedded inside PbTe matrix, (b) High-resolution image (HRTEM) of 1% CdTe sample showing the coexistence of domains with different features, (c) The fast Fourier transform (FFT) of the matrix domain, and (d) The fast Fourier transform of the nanoparticle domain.

CONCLUSIONS

Thermoelectric properties of PbTe-x% CdTe samples were investigated in detail in an effort to test the theoretical prediction that resonance levels might increase the Seebeck coefficient and ZT. No significant enhancement in the Seebeck coefficient was observed. However, the high resolution transmission electron microscopy study shows that nanostructuring plays a role in reducing the lattice thermal conductivity. A maximum ZT of ~1.2 at ~720 K was achieved for the 1% CdTe sample.

ACKNOWLEDGMENTS

Financial support from the Office of Naval Research is gratefully acknowledged.

REFERENCES

1. L. E. Bell, Science **321**, 1457 (2008).
2. G. J. Snyder and E. S. Toberer, Nature Mater. **7**, 105 (2008).
3. K. F. Hsu, S. Loo, F. Guo, W. Chen, J. S. Dyck, C. Uher, T. Hogan, E. K. Polychroniadis, and M. G. Kanatzidis, Science **303**, 818 (2004).
4. J. Androulakis, C.-H. Lin, H.-J. Kong, C. Uher, C.-I Wu, T. Hogan, B. A. Cook, T. Caillat, K. M. Paraskevopoulos, and M. G. Kanatzidis, J. Am. Chem. Soc. **129**, 9780 (2007).
5. J. R. Sootsman, H. Kong, C. Uher, J. J. D'Angelo, C.-I Wu, T. P. Hogan, T. Caillat, and M. G. Kanatzidis, Angew. Chem. Int. Ed. **47**, 8618 (2008).
6. S. Ahmad, S. D. Mahanti, K. Hoang, and M. G. Kanatzidis, Phys. Rev. B **74**, 155205 (2006).
7. J. P. Heremans, V. Jovovic, E. S. Toberer, A. Saramat, K. Kurosaki, A. Charoenphakdee, S. Yamanaka, G. J. Snyder, Science, **321**, 554 (2008).
8. V. Leute and R. Schmidt, Z. Phys. Chem. **172**, 81 (1991).
9. S. J. Thiagarajan, V. Jovovic, and J. P. Heremans, Phys. Stat. Sol. (RRL) **1**, 256 (2007).
10. B. Poudel, Q. Hao, Y. Ma, Y. Lan, A. Minnich, B. Yu, X. Yan, D. Wang, A. Muto, D. Vashaee, X. Chen, J. Liu, M. S. Dresselhaus, G. Chen, and Z. Ren, Science **320**, 634 (2008).

Mater. Res. Soc. Symp. Proc. Vol. 1166 © 2009 Materials Research Society   1166-N08-01

Thermal Conductivity Reduction Paths in Thermoelectric Materials

C. Godart [1], A.P. Gonçalves [2], E.B. Lopes [2], B. Villeroy [1]
[1]CNRS, ICMPE, CMTR, 2/8 rue Henri Dunant, 94320 Thiais, France
[2]Dep. Química, Instituto Tecnológico e Nuclear/CFMC-UL, P-2686-953 Sacavém, Portugal

ABSTRACT

The figure of merit ZT = $\sigma S^2 T/\kappa$ (S the Seebeck coefficient, σ and κ the electrical and thermal conductivity respectively) is an essential element of the efficiency of a thermoelectric material for applications which convert heat to electricity or, conversely, electric current to cooling. From the expression of the power factor σS^2 it was deduced that a highly degenerated semiconductor is necessary. In order to reduce the lattice part of the thermal conductivity, various mechanisms were tested in new thermoelectric materials and those had been the topics of several reviews. These include cage-like materials, effects of vacancies, solid solutions, complex structures (cluster, tunnel, ...,), micro- and nano-structured systems, and more recently semiconducting glasses. We plan to review such aspects in the modern thermoelectric materials and include results of the very last years. Moreover, as micro- and nano-composites seem to be promising to increase ZT in large size samples, we will also briefly discuss the interest of spark plasma sintering technique to preserve the micro- or nano- structure in highly densified samples.

INTRODUCTION

Thermoelectric (TE) effects include the transformation of caloric energy to electric energy or its reverse, and their applications consequently include the two aspects: (micro)cooling or electricity generation from heat sources. However, the efficiency was not sufficient to compete with the cooling by compression/expansion cycles or for economically profitable electricity production.

For both cooling or electricity generation, a TE module is made of couples, each couple includes a p-type material (S>0) and a n-type material (S<0) in which charge carriers are holes and electrons, respectively, with S the Seebeck coefficient. These two materials are joined by a conducting material assumed to be with S=0. The two branches of the couple and all other couples of the module are connected electrically in series and thermally in parallel. In order to improve the efficiency of the conversion, it is necessary to optimize the transport properties of the materials, in order to maximize the so-called figure of merit:

$$ZT=\sigma S^2 T/\lambda \qquad \text{Eq. 1}$$

with $\sigma=1/\rho$ and λ the electrical and thermal conductivities, respectively.

The expression of the figure of merit clearly sums up the difficulty to optimize the materials. Intuitively, it seems difficult for a material to simultaneously possesses a good electrical conductivity, characteristic of metals, and a bad thermal conductivity, characteristic of an insulator. Closer examination of the power factor σS^2 leads to consider semiconductors with a small gap and a carrier concentration in the range [10^{18}-10^{21} cm^{-3}]. The second parameter is the thermal conductivity which, in such semiconductors, must be reduced.

Intuitively, a good thermal conductivity would prevent the establishment of a large temperature gradient. However, the thermal conductivity includes mainly two components: an electronic contribution λ_e, due to the transport by carriers, and a lattice contribution due to phonons λ_L: $\lambda=\lambda_e+\lambda_L$. The electronic part of the thermal conductivity is related to the electronic conductivity (hold by the power factor) via the Wiedemann-Franz law:

$$\lambda_e=L_0T\sigma, \qquad\qquad\qquad Eq. 2$$

with L_0 the Lorentz factor. In metals, it is equal to the Lorentz number:

$L_0 = \dfrac{\pi^2}{3}\left(\dfrac{k}{e}\right)^2 = 2,45\cdot10^{-8}$ $V^2.K^{-2}$, value generally used as a first approximation for the TE

semiconductors. Replacing λ by its two components and with the Wiedemann-Franz, leads to:

$$ZT = \frac{S^2T\sigma}{LT\sigma+\lambda} \text{ or } ZT = \frac{S^2}{L}\frac{\lambda}{\lambda_e+\lambda} . \qquad\qquad Eq. 3$$

This last expression shows that the optimization of ZT implies the minimization of the phonons contribution to λ. Assuming a negligible lattice contribution, from $ZT=S^2/L$, a ZT of 1 or 3 would be obtained from a Seebeck coefficient of 156 $\mu V/K$ or 270 $\mu V/K$, respectively. However, the minimization should not modify the electrical conductivity (as ZT increases with λ_e/λ_L).

For both cooling or profitable electricity generation,, new concepts [1] lead, from ~1995, to rather remarkable progress.

The two main research directions happen to be - the development of new materials with complex or open structures, and - the development of already known materials under new low dimensionality forms (quantum wells, nano-wires, nanograins, thin films, nano-composites, ..). Among new materials having interesting TE characteristics for electricity generation from heat sources in the range 200-800°C we will emphasize the role of the structure and the potential of micro- and nano-structures to exceed the 14% efficiency already reported by [2].

Moreover, as micro- and nano-composites seem to be promising to increase ZT in large size samples, we will also briefly discuss the interest of spark plasma sintering technique which has been extensively used, since around 2005, to preserve the micro- or nano- structure in highly densified TE samples.

BULK MATERIALS

In the early 90's, Slack propose [1] to look for materials which conducts electricity as a crystal and heat as a glass, so-called "Phonon Glass Electron Crystal" (PGEC). This implies to find selective diffusion processes that affect more the phonons than the carriers.

Various physical processes inducing significant phonon diffusion have been tested on TE materials, let us quote those that are somehow related with the structure:

- a complex crystalline structure will increase the number of optical phonon modes, whereas the heat is mainly conducted by the acoustic phonons;

- the insertion of heavy atoms in empty cages of the crystalline structure may let them rattle independently of the lattice and so, create new phonon modes (weakly dispersive optical modes);

- solid solutions, between different materials with the same structure, increase the disorder and create an important phonon diffusion by mass fluctuations on the sites [Abeles 1963], such mass fluctuations can also be obtained by vacancy creation in the material;

-impurities and point defects will diffuse phonons [3, 4, 5]: this leads to study micro- or nano-composites materials (mixing of a good TE material with another neutral for TE), or use "exotic" synthesis or shaping techniques thus inducing large amount of such defects,

- grain boundaries will affect phonons (and also electrical conductivity), this leads to study nano-crystalline materials to increase the number of such boundaries or even to reduce the mean free path of phonons (when the sizes of the nano-grains are comparable with that of the mean free path of phonons).

Those different processes are not mutually exclusive and some of them can be produced simultaneously in the same TE material.

From 1960, all currently used TE materials (see Table 1) were already known and their performances, bound to a stagnant ZT ~1, have not changed much up to 1990.

The main commercial applications, were based on Bi_2Te_3- type of material that is mainly used for cooling. Let us remark that no material was efficient in the temperature range 400-700K. For higher temperatures, PbTe and TAGS (Te-Ag-Ge-Sb) have ZT ~1 in n- and p- type respectively.

	Bi-Sb	Bi_2Te_3- Sb_2Te_3	$(Bi,Sb)_2 (Te,Sb)_3$	---	PbTe	Te-Ag-Ge-Sb	---	Si-Ge
Type	n	n,p	n,p	-	n	p	-	n,p
T_u (K)	200	<300	~300-400	-	700	750	-	1000
ZT at T_u	1.1 (H)	0.8	0.9	-	0.8	1.1	-	0.6

Table 1. ZT of conventional materials at their optimal temperature of use T_u (H means a value obtained under a magnetic field), --- temperature ranges without good TE materials.

In solids, the atoms displacements are frozen only at zero Kelvin temperature. At finite temperatures, the atoms vibrate around their equilibrium position on their crystallographic site, the mean square displacement amplitude of such atoms being called the Atomic Displacement Parameter, ADP, which is useful to estimate the minimum value of the lattice contribution to the thermal conductivity and is thus of interest in TE materials [6].

The minimal value of the lattice part of the thermal conductivity is given by $\lambda_L=(1/3)C_v v_s d$, with v_s the speed of sound, d the phonon mean free path and C_v the specific heat estimated from the Dulong and Petit law

$$C \xrightarrow[T \to \infty]{} 3R=3N_A k_B \qquad \text{Eq. 4}$$

(with R the constant of gas, N_A the Avogadro number). In the case of a cage like material, like LaB_6 (not a TE material), (Fig. 1), the specific heat has been calculated from ADP of La and B atoms.

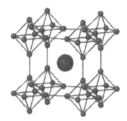

Fig. 1. Cage like structure of LaB_6, the La atom ● is surrounded by B ● groups.

The La atom is considered as an Einstein oscillator in a Debye solid of B-atoms. $C_V = f.C_D + ((1-f)C_E$, with f the fraction of atoms contributing to the solid of Debye, (1-f) that of Einstein oscillators:

$$C_D = 9N_A k_B (\frac{T}{\theta_D})^3 \int_0^{\theta_D/T} \frac{x^4 e^x}{(e^x-1)^2} dx \qquad \text{Eq. 5}$$

$$C_E = 3N_A k_B (\frac{\theta_E}{T})^2 \frac{e^{\theta_E/T}}{(1-e^{\theta_E/T})^2} \qquad \text{Eq. 6}$$

where θ_D and θ_E represents the Debye and Einstein temperatures respectively

Cage like TE materials have a rigid sub-lattice responsible for the electrical conductivity and large empty cages. When the cages are filled with heavy atoms, these atoms, weakly bound to the cage, can vibrate inside with large amplitudes of vibrations. Initially it has been proposed that these vibrations are incoherent ("rattling") and act as traps for the acoustic phonons and consequently decrease the thermal conductivity. According to more recent works [7], [8], the vibrations of the inserted atom are optical phonons (coherent vibrations) mainly without dispersion (localized character) and with a weak energy which interfere strongly with the acoustic phonons and therefore decrease the thermal conductivity.

The two most studied families of cage like TE materials are the filled skutterudites $A_y M_4 X_{12}$ (A: an electropositive element, M a metal and X=P, As, Sb) and the clathrates of the type $A_8 Y_{16} X_{30}$ (X and Y being actually mainly Ga and Ge), both of them leading to increased values of ZT. Many TE series of materials have been studied during the past 15 years and few of them not only lead to increased values of ZT by about 50% (ZT~1.5) but also to an increase of the temperature ranges where more efficient TE materials now exist (as compared to Table 1), see Figures 2, 3.

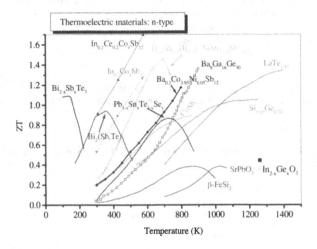

In the n-type series, cage like materials (skutterudites, clathrates), complex substitutions (half-Heusler), disordered phase ($LaTe_{1+x}$) lead to better ZT values.

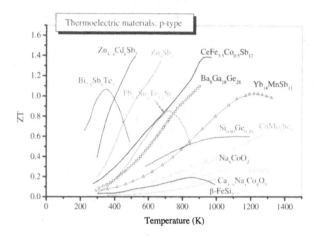

Figure 3: Best p-type TE bulk materials (non exhaustive)

In the p-type series, cage like materials (skutterudites, clathrates), complex structures ($Yb_{14}MnSb_{11}$, Chevrel phases), disordered phases (Zn_4Sb_3) lead to better ZT values.

In mainly all n- and p-type of these new series of materials, vacancies contribute to the highest ZT.

Many other structural families may have empty cages in which the insertion of atoms with various oxidation degrees may lead to a semiconducting state, a prerequisite for good TE properties.

Some very recent reports show that new possibilities exist:

In the phase with Mo_9Se_{11} clusters, the structure owns tunnels, and the insertion of atoms (Ag, Cs, Cl) in cages /[9]/ leads to semiconducting or semi-metallic compounds, depending on the number of electrons per cluster. A Seebeck coefficient of $72\mu V/K$ and a weak resistivity have been reported in $Ag_{3.6}Mo_9Se_{11}$ /[10]/.

A careful analysis of the main general approaches recently developed to obtain improved thermoelectric systems shows that the new materials must have complex structures, allow the presence of inclusions and impurities, and should have mass fluctuations and disorder. A type of materials that follows all these principles are the glasses /[11]/. Indeed, they have extremely complex structures, with a certain degree of order only at small distances, and can present mass fluctuations, easily allowing high concentrations of inclusions and impurities. The reported electrical conductivity and Seebeck coefficient values of metallic glasses indicate that they are not suitable for thermoelectric applications, as their Seebeck coefficients values are typical of metals (+ -5μV/K). To identify glasses with improved thermoelectric performances it is necessary to study low gap semiconductor and semimetal glasses. In particular, $Ge_{20}Te_{80}$, is based only on two elements, is mainly formed by heavy atoms, has high Seebeck coefficients, albeit presenting small electrical conductivity values. However, doping it with copper or silver can dramatically increase the electrical conductivity with a small decrease of the Seebeck coefficient, which lead to a huge increase of the power factor: the $Ge_{20}Te_{80}$ presents a value of $3x10^{-3}$ μW/K^{2}m, four orders of magnitude lower than the $Cu_{27.5}Ge_{2.5}Te_{70}$, which has a value higher than 60 μW/K^{2}m. The very low values of thermal conductivity observed in this type of glasses (<0.2 W/Km for $Ge_{20}Te_{80}$) lead to ZT >0.1 at 300 K for the $Cu_{27.5}Ge_{2.5}Te_{70}$ sample/[12]. Other substitutions have been recently reported in amorphous $Ge_{20}Te_{80-x}Se_x$ /[13]/ making them interesting materials for thermoelectric properties.

The best ZT values in oxides reach 0.4 for the p-type and 0.3 for the n-type in conventional oxides and of 0.55 in misfits oxides with complex structure. The structure of In_2O_3 (space group Ia-3) shows alternatively ordered and disordered In-planes. The solubility limit of Ge is no more than 0.5 atom%, for higher rate of Ge the material forms a composite In_2O_3 + inclusions of micronic size (~100μm) of $In_2Ge_2O_7$. In that composite, the thermal conductivity is notably reduced and a ZT of 0.45 at 1243K has been observed which is actually the best value for a n-type oxide.

This shows the potential of micro-composites, as well as nano-composites (see below), to improve ZT values as compared to the "parent" materials.

NANO MATERIALS

Transport properties in micro-, and nano- structures differ from that in bulk 3D (three-dimensional) materials. One of the original ideas was to increase the electronic density of states through electron confinement /[14]/. This is illustrated in the figure 4 from Dresselhaus /[15]/.

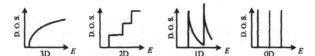

Figure 4: Electronic density of states versus dimensionality from /[15]/.

The thermal conductivity of nano-structures like super-lattices (material with periodic change of nano-metric layers of various elements or substances) is weaker than in bulk materials. This is a positive aspect for TE properties. However, these nano-structures are generally obtained by thin films techniques (Molecular Beam Epitaxy, Knudsen cells, Pulsed laser, CVD, ..), too costly to be used for the mass production of materials necessary for large scale applications in electricity generation. Moreover, their weak thickness is not very compatible with the formation of large temperature gradients. Conversely, they are well adapted to micro-cooling in electronics.

Electronic transport properties are strongly modified by band structure dimensionality effects /[16]/. To summarize, in a 3D material, S, σ and λ are bound and it is difficult to optimize both of them, in the case of lower dimensionality new possibilities exist to independently adjust them. In addition, new interfaces are created which can increase the phonon diffusion, more than that of carriers, leading to an increase value of ZT.

It is remarkable that the improvement of ZT has been theoretically predicted /[14], [17]/ (see figure 5), before its experimental observation.

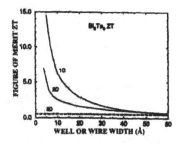

Figure 5: Evolution of ZT with the dimensionality (from /[17]/).

The experimental proofs, in various forms of nano-structures, concern, for instance:
- super lattices including quantum wells PbTe-PbSeTe /[18]/, in which ZT value of 1.6 has been calculated (in front of 0.4 for the bulk)
- Enhanced thermoelectric performance in PbTe-based superlattice structures/[19]/ (~double ZT for quantum dots),
- super lattices of Bi_2Te_3/Sb_2Te_3, with a ZT of 2.4 (in front of 1.0 for the bulk) /[20]/,
- a decrease of the thermal conductivity of a Ge matrix by inclusions of Si-nano-wires which depends on the diameter of the nano-wires /[21]/,
- an increase of the ZT value in various compositions of Bi_2Te_3 by addition of Bi_2Te_3-nano-powders /[22]/, ...

The series of cubic materials $AgPb_mSbTe_{2+m}$ has lead to type n semiconductor with high ZT values when m=10 (ZT=1 at 700K) or m=18 (ZT=2.2 at 800K) /[23]/. Initially believed to be a bulk property, it was assumed from high resolution transmission microscopy that Ag-Sb atoms arrangements lead to an heterogeneous nano-structure material /[24]/, that has also been described as nano-clusters of $AgPb_3SbTe_5$ in PbTe matrix /[25]/. Such nano-phases could be responsible for high ZT values recently reported in various new systems (derived from conventional TE materials) like: ZT=1.7 at 650K in $Na_{0.95}Pb_{19}SbTe_{22}$ /[26]/, ZT = 1.5 at 640K in $(Pb_{0.95}Sn_{0.05}Te)_{0.92}(PbS)_{0.08}$ /[27]/, ZT=1.45 at 625K in $Ag_{0.5}Pb_6Sn_2Sb_{0.2}Te_{10}$ //[28], also ZT = 1.5 in $(GeTe)_x(AgSbTe_2)_{100-x}$ /[29]/, micro-inclusions were already observed long ago in $AgSbTe_2$ /[30]/. The difficult control of the nature, size and dispersion of these nano-phases could be responsible of the mismatch of literature results. The effects of annealing and of eventual temperature cycling which could induce a growth of these nano-objects have not yet been studied.

SHAPING WITH SPS

This last example, as well as the case of In_2O_3+Ge, show that the study of micro- and nano-composites may lead to further improvement of TE performances due to size effects. However, samples of large sizes (> centimeter) are requested for large scale applications. The shaping of such powder (or powdered) samples can be realized by hot pressing techniques. In the recent studies, this shaping is generally obtained by Spark Plasma Sintering (SPS) technique, a technique of fast sintering under pressure, that mainly avoids the grain growth observed with the conventional hot press. A small size SPS machine, as we acquired in CNRS Thiais (Syntex 515S), is able to perform sintering under conditions up to 100MPa and 2000°C, with non costly graphite crucibles. Sample diameters of up to 20-25 mm are obtained for sintering temperature up to ~1500°C for ceramics and ~1000°C for intermetallics; for higher temperatures smaller diameter can be produced. These temperatures are generally sufficient as the pressure helps the sintering process. Pressures of up to 500MPa have been realized with other crucibles (W-C) at lower temperatures and diameter.

Such SPS technique has been used with various TE conventional powders (grains smaller than a few 10μm) of Bi_2Te_3 [31], skutterudites [32, 33], clathrates [34], [35], half Heusler [36], Zn_4Sb_3 [37], doped-Mg_2Si [38]/or functionally graded Mg_2Si-$FeSi_2$ [39], PbTe and TAGS [40], PbTe [41], ZnO derivatives [42], etc..

Of course, the nanostructure, essentially preserved by the SPS, should not be destroyed during the temperature cycling for electricity generation at high temperatures. For materials, like Bi_2Te_3 to be used around room temperature the problem is certainly not that important. However, for skutterudites prepared by mechanical alloying and sintered by SPS [43, 33], we observe by Thermal Differential Analysis that grain growth occurs during the first heating at ~350°C [43]. In such a case, a solution will be to make nano-composites in which a nano-object is inserted in an already good TE matrix, the chemistry of the nano-objects being chosen to not react with the matrix at the temperature of use of the thermo-element.

Recently, the technique has been successfully applied in various composite systems, the starting nano-particles being produced by sol-gel method, electrochemistry,... but more often from ball milling:
- various compositions of Bi_2Te_3 with addition of Bi_2Te_3-nano-powders [44]
- improved ZT in nano-composite (grain sizes ~100-150nm) of PbTe [45]
- synthesis of $Yb_yCo_4Sb_{12}/Yb_2O_3$ composites with decreased thermal conductivity [46]
- improved ZT of composite made of nano-particles of SiC dispersed in Bi_2Te_3 [47],
- optimization of p-type segmented $FeSi_2/Bi_2Te_3$ thermoelectric material [48] with an experimental power of ~600W/m^2 for a hot temperature of ~500°C,
- shaping of binary skutterudites with C_{60} inclusions [49]
- mixing of micro-nano- skutterudites [50]
- Bi addition in Heusler V_2FeAl [51]
- nano-inclusions of ZrO_2 in the half Heusler $Zr_{0.5}Hf_{0.5}Ni_{0.8}Pd_{0.2}Sn_{0.99}Sb_{0.01}$ [52]

Moreover, others properties are sometimes improved by the use of SPS, for instance:
- SPS is well adapted to prepare various kinds of functionally graded materials (by sintering stacked powders),
- better mechanical properties of Bi_2Te_3 wafer films (down to 100μm thickness) [53],
- electric contact or protection against reactivity with electrode can be made, in few cases, by direct sintering of a thin metal foil with the powder, for instance joining skutterudite to Mo electrode via a Ti intermediate layer [54, 55]
- use of Ta foil to prevent reaction of graphite with TE oxide
- magnetic alignment of cobaltites along c-axis to improve ZT [56],

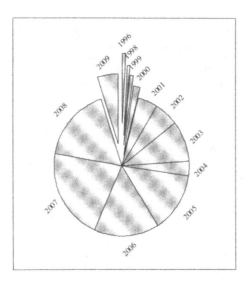

Figure 6: Annual percentage of publications on TE materials using SPS (Ref.: 259 publications in April 2009 from WEB of knowledge, areas are proportional to percentage).

It is quite remarkable that such sintering processes take place within a short period of time (typically few minutes) and can lead to densities higher than 95% of the theoretical X-ray density, a very crucial parameter for TE transport properties. Moreover, improved power factors or ZT values are generally reported. This explains the increasing use of this technique for the synthesis and/or shaping of TE materials and micro-, nano-composites, since 2005 (see Fig. 6).

The forthcoming years should prove to be crucial for the understanding of transport mechanisms in nano-composites and consequently to take advantage of the beneficial effects that such systems can have on TE properties as compared to the detrimental effects like, for instance, the decrease of the electrical conductivity. Such understanding could help to find new semiconducting systems in which micro-, nano-inclusions develop and are stable. Improvement of final densities of the materials during sintering by SPS will contribute to increase the electrical conductivity of these materials and should therefore be beneficial for TE properties.

ACKNOWLEDGMENTS

This work was supported by the bi-lateral French-Portuguese GRICES 2007-2008 program and European COST P16 program. One of us (CG) acknowledges CNRS support to its participation in "Ile de France" SPS project in 2007.

conventional hot press. A small size SPS machine, as we acquired in CNRS Thiais (Syntex 515S), is able to perform sintering under conditions up to 100MPa and 2000°C, with non costly graphite crucibles. Sample diameters of up to 20-25 mm are obtained for sintering temperature up to ~1500°C for ceramics and ~1000°C for intermetallics; for higher temperatures smaller diameter can be produced. These temperatures are generally sufficient as the pressure helps the sintering process. Pressures of up to 500MPa have been realized with other crucibles (W-C) at lower temperatures and diameter.

Such SPS technique has been used with various TE conventional powders (grains smaller than a few 10μm) of Bi_2Te_3 [31], skutterudites [32, 33], clathrates [34], [35], half Heusler [36], Zn_4Sb_3 [37], doped-Mg_2Si [38]/or functionally graded Mg_2Si-$FeSi_2$ [39], PbTe and TAGS [40], PbTe [41], ZnO derivatives [42], etc..

Of course, the nanostructure, essentially preserved by the SPS, should not be destroyed during the temperature cycling for electricity generation at high temperatures. For materials, like Bi_2Te_3 to be used around room temperature the problem is certainly not that important. However, for skutterudites prepared by mechanical alloying and sintered by SPS [43, 33], we observe by Thermal Differential Analysis that grain growth occurs during the first heating at ~350°C [43]. In such a case, a solution will be to make nano-composites in which a nano-object is inserted in an already good TE matrix, the chemistry of the nano-objects being chosen to not react with the matrix at the temperature of use of the thermo-element.

Recently, the technique has been successfully applied in various composite systems, the starting nano-particles being produced by sol-gel method, electrochemistry,... but more often from ball milling:
 - various compositions of Bi_2Te_3 with addition of Bi_2Te_3-nano-powders [44]
 - improved ZT in nano-composite (grain sizes ~100-150nm) of PbTe [45]
 - synthesis of $Yb_yCo_4Sb_{12}$/Yb_2O_3 composites with decreased thermal conductivity [46]
 - improved ZT of composite made of nano-particles of SiC dispersed in Bi_2Te_3 [47],
 - optimization of p-type segmented $FeSi_2$/Bi_2Te_3 thermoelectric material [48] with an experimental power of ~600W/m² for a hot temperature of ~500°C,
 - shaping of binary skutterudites with C_{60} inclusions [49]
 - mixing of micro-nano- skutterudites [50]
 - Bi addition in Heusler V_2FeAl [51]
 - nano-inclusions of ZrO_2 in the half Heusler $Zr_{0.5}Hf_{0.5}Ni_{0.8}Pd_{0.2}Sn_{0.99}Sb_{0.01}$ [52]

Moreover, others properties are sometimes improved by the use of SPS, for instance:
 - SPS is well adapted to prepare various kinds of functionally graded materials (by sintering stacked powders),
 - better mechanical properties of Bi_2Te_3 wafer films (down to 100μm thickness) [53],
 - electric contact or protection against reactivity with electrode can be made, in few cases, by direct sintering of a thin metal foil with the powder, for instance joining skutterudite to Mo electrode via a Ti intermediate layer [54, 55]
 - use of Ta foil to prevent reaction of graphite with TE oxide
 - magnetic alignment of cobaltites along c-axis to improve ZT [56],

DISCUSSION

25   H. Lin, E. S. Bozin, S. J. L. Billinge, E. Quarez, and M. G. Kanatzidis, Phys. Rev. B **72**, 174113 (2005).

26   P. F. P. Poudeu, J. D'Angelo, A. D. Downey, J. L. Short, T. P. Hogan, and M. G. Kanatzidis, Angew. Chem., Int. Ed. Engl. **45**, 3835 (2006).

27   J. Androulakis, C. H. Lin, H. J. Kong, C. Uher, C. I. Wu, T. Hogan, B. A. Cook, T. Caillat, K. M. Paraskevopoulos, and M. G. Kanatzidis, J. Am. Chem. Soc. **129**, 9780 (2007).

28   J. Androulakis, K. F. Hsu, R. Pcionek, H. Kong, C. Uher, J. J. D'Angelo, A. Downey, T. Hogan, and M. G. Kanatzidis, Advanced Materials **19**, 1170 (2006).

29   S. H. Yang, T. J. Zhu, T. Sun, J. He, S. N. Zhang, and X. B. Zhao, Nanotechnology **19**, 245707 (2008).

30   R. W. Armstrong, J. Faust, J.W., and W. A. Tiller, J. Appl. Phys. **31**, 1954 (1960).

31   D. Kusano and Y. Hori, J. Jpn. Inst. Met. **66**, 1063 (2002).

32   L. D. Chen, X. Shi, X. Y. Huang, and S. Q. Bai, *23rd International Conference on Thermoelectrics- Adelaide, Australia 25 – 29 July 2004*, (2004).

33   C. Recknagel, N. Reinfried, P. Höhn, W. Schnelle, H. Rosner, Y. N. Grin, and A. Leithe-Jasper, Science and Technology of Advanced Materials **8**, 357 (2007).

34   H. Anno, M. Hokazono, H. Takakura, and K. Matsubara, *24th International Conference on Thermoelectrics- ICT2005*, (2005).

35   M. Hokazono, H. Anno, and K. Matsubara, Mater. Trans. JIM **46**, 1485 (2005).

36   X. Y. Huang, Z. Xu, L. D. Chen, and X. F. Tang, Key Eng. Mater. **240**, 79 (2003).

37   T. Souma, G. Nakamoto, and M. Kurisu, *22nd International Conference on Thermoelectrics - ICT*, (2003), p. 279.

38   J. Tani and G. Kido, Physica B **364**, 218 (2005).

39   A. Sugiyama, K. Kobayashi, K. Ozaki, T. Nishio, and A. Matsumoto, J. Jpn. Inst. Met. **62**, 1082 (1998).

40   Y. Noda, K. Mizuno, Y. S. Kang, M. Niino, and I. A. Nishida, J. Jpn. Inst. Met. **63**, 1448 (1999).

41   S. Yoneda, E. Ohta, H. T. Kaibe, I. J. Ohsugi, I. Shiota, and I. A. Nishida, Mater. Trans. JIM **42**, 329 (2001).

42   K. H. Kim, S. H. Shim, K. B. Shim, K. Niihara, and J. Hojo, J. Am. Ceram. Soc. **88**, 628 (2005).

43   D. Bérardan, E. Alleno, C. Godart, H. Benyakoub, H. Flandorfer, O. Rouleau, and E. Leroy, *24th International Conference on Thermoelectrics ICT2006,- Vienna, Austria- 6-10 Aug.*, (2006), p. 151.

44   H. L. Ni, X. B. Zhao, G. Karpinski, and E. Müller, J. Mater. Sci. **40**, 605 (2005).

45   J. Martin, G. S. Nolas, W. Zhang, and L. Chen, Appl. Phys. Lett. **90**, 222112 (2007).

46   X. Y. Zhao, X. Shi, L. D. Chen, W. Q. Zhang, S. Q. Bai, Y. Z. Pei, and X. Y. Li, Appl. Phys. Lett. **89**, 092121 (2006).

47   J. F. Li and J. Liu, Phys. Status Solidi A **203**, 3768 (2006).

48   J. L. Cui, Mater. Lett. **57**, 4074 (2003).

49   X. Shi, L. D. Chen, S. Q. Bai, X. Y. Huang, and X. F. Tang, *21st International Conference on Thermoelectrics- Long Beach, California, USA - Aug. 25-29*, (2002), p. 68.

50   J. X. Zhang, Q. M. Lu, X. Zhang, and Q. Wei, *24th International Conference on Thermoelectrics ICT2006,- Vienna, Austria- 6-10 Aug.*, (2006), p. 148.

[51]   M. Mikami and K. Kobayashi, J. Alloys Compd. **466**, 530 (2008).
[52]   L. D. Chen, X. Y. Huang, M. Zhou, X. Shi, and W. B. Zhang, J. Appl. Phys. **99**, 064305 (2006).
[53]   H. Böttner, D. G. Ebling, A. Jacquot, J. König, L. Kirste, and J. Schmidt, Physica Status Solidi RRL **1**, 235 (2007).
[54]   J. F. Fan, L. D. Chen, S. Q. Bai, and X. Shi, Mater. Lett. **58**, 3876 (2004).
[55]   D. Zhao, X. Li, L. He, W. Jiang, and L. Chen, Intermetallics **17**, 136 (2009).
[56]   S. Horii, I. Matsubara, M. Sano, K. Fujie, M. Suzuki, R. Funahashi, M. Shikano, W. Shin, N. Murayama, J. Shimoyama, and K. Kishio, Jpn. J. Appl. Phys., Part 1 **42**, 7018 (2003).

Oxide Materials

Mater. Res. Soc. Symp. Proc. Vol. 1166 © 2009 Materials Research Society   1166-N03-13

Preparation of Delafossite CuYO$_2$ by Metal-Citric Acid Complex Decomposition Method

Keishi Nishio[1], Tomomi Okada[1], Naoto Kikuchi[2], Satoshi Mikusu[3], Tsutomu Iida[1], Kazuyasu Tokiwa[3], Tsuneo Watanabe[3] and Tohru Kineri[4]

[1]Material Science and Technology, Tokyo University of Science, 2641 Yamzaki, Noda-shi, Chiba 278-8510, Japan
[2]Nanoelectronics Research Institute, National Institute of Advanced Industrial Science and Technology, 1-1-1 Umezono, Tsukuba, Ibaraki 305-8568, Japan
[3]Department of Applied Electronics, Tokyo University of Science, 2641 Yamzaki, Noda-shi, Chiba 278-8510, Japan
[4]Department of Materials Science & Environmental Engineering, Tokyo University of Science, Yamaguchi, 1-1-1 Daigaku-doori, Sanyo-Onoda-shi, Yamaguchi, 756-0884, Japan

ABSTRACT

Delafossite CuYO$_2$ and Ca doped CuYO$_2$ were prepared by thermal decomposition of a metal-citric acid complex. The starting solution consisted of Cu acetate, Y acetate and Ca acetate as the raw materials. Citric acid was used as the chelating agent, and acetic acid and distilled water were mixed as a solvent. The starting solutions were heated at 723 K for 5 h after drying at 353 K. The obtained powders were amorphous and single phase of orthorhombic Cu$_2$Y$_2$O$_5$ was obtained by heat-treated the amorphous powder at a temperature range between 1073 and 1373 K for 3 h in air. Furthermore, Heat-treating the obtained orthorhombic Cu$_2$Y$_2$O$_5$ at above 1373 K in air caused it to decompose into Y$_2$O$_3$, CuO and Cu$_2$O. On the other hand, the sample powder prepared from a starting solution without citric acid, i.e., single phase of orthorhombic Cu$_2$Y$_2$O$_5$ could not be obtained under the same synthesis conditions as that for a solution with citric acid. We were able to obtain delafossite CuYO$_2$ and Ca doped CuYO$_2$ from orthorhombic Cu$_2$Y$_2$O$_5$ under a low O$_2$ pressure atmosphere at above 1223 K. The obtained delafossite CuYO$_2$ composed hexagonal and rhombohedral phases. The color of the CuYO$_2$ powder was light brown and that of Ca-doped CuYO$_2$ was light green. Diffraction peaks in the XRD pattern were slightly shifted by doping Ca for CuYO$_2$, and these peaks shifted toward to a high diffraction angle with an increasing amount of doped Ca. From these results, we concluded that Ca doped delafossite CuYO$_2$ could be obtained by thermal decomposition of a metal-citric acid complex.

INTRODUCTION

Thermoelectric materials having the properties of large thermo-power, low resistivity, and low thermo-conductivity have been investigated. Recently, many reseachers have reported p-type semiconductors with layered metal oxides, i.e., Na-Co-O and Ca-Co-O, because they show low resistivity and low thermo-conductivity [1-3]. For example, Ca$_3$Co$_4$O$_9$ consists of Ca$_2$CoO$_3$ and CoO$_2$ blocks alternately stacked along the c-axis to form a layered structure. Thus, the physical properties are highly two-dimensional in an a-b plane compared with those in the c-axis direction. This indicates very high thermoelectric properties. These layered metal oxide compounds promising candidate for thermoelectric application at high temperature in air. It has been reported that $CuYO_2$ with a delafossite structure is a p-type, wide-band gap semiconductor with a high Seebeck coefficient of $+274\mu VK^{-1}$ that has the same structure as the mineral delafossite $CuFeO_2$ [4]. The delafossite structures (i.e. hexagonal and rhombohedral) belong to a layered structure [5]. The crystal structure of delafossite $CuMO_2$ (M: metal) oxides is composed of alternatively stacked of Cu cation layers and MO_2 layers along the c-axis. There is no oxygen within the Cu^+ layer and only two oxygen atoms are linearly coordinated to each Cu cation (O-Cu-O dumbbell) layer along the c-axis. MO_6 octahedras, which share edges with each other, constitute the MO_2 layer. The closed-packed Cu^+ layer perpendicular to the c-axis is well known to act as a hole conduction path. Moreover, delafossite materials have a large σ value with low carrier concentration and high Seebeck coefficient. Synthesizing delafossite $CuYO_2$, however, requires high temperature heat treatment under a low O_2 pressure atmosphere for a considerable long of time. In this study, we tried to prepare a single phase of p-type $CuYO_2$ semiconductor for application to thermoelectric devices to control carrier concentration by doping the $CuYO_2$ with Ca, and to achieve a lower temperature range and a shorter time for the process [6-8]. The Pechini's method [9] is very famous as a simple method for preparing metal oxide powders where polymeric precursors are made from metal salts, ethylene glycol, and citric acid by low-temperature heat-treatment. This method allows the metal cations to be mixed at the molecular level and stoichiometric compositions to be achieved by chelating the metal ions in solution by citric acid, on which many studies have been reported [10-14]. Furthermore, the process offers several advantages in the fabrication of ceramic thin films, including low cost, homogeneous compositions, high purity, and low heat-treatment temperatures. In this study, we tried to obtain $CuYO_2$ and Ca doped $CuYO_2$ powders from a metal–citric acid complex for use as a non-polymeric precursor because polymeric precursors generate CO_2 gas when they burn out.

EXPERIMENT

Metal salts ($Y(OCOCH_3)_3$, $Cu(OCOCH_3)_2$, and $Ca(OCOCH_3)_2$) were used as raw materials. These metal salts were dissolved in a solution that comprised citric acid dissolved in dilute acetic acid at room temperature. To obtain a precursor powder, the precursor solution was calcined at several temperatures for 3 h after drying at 353 K and 723 K in air. The precursor powder was then heat-treated in N_2 gas. To shape the powder into a green body, the precursor powder was placed in a die and subjected to uni-axis pressure of 500 MPa. After the green body was formed, it was sintered at 1273K for 20 h in Ar gas atmosphere. Several green bodies were made as samples in this way.

The relative densities of the samples were measured using Archimedes' method. A scanning electron microscope (SEM) investigation of them was performed using an SEM–5800LV system (JEOL Ltd., Japan). The XRD patterns of powders and ceramics were measured using an X'Pert-Pro MDR system (Spectris Co.) or an Ultima IV system (Rigaku Co., Japan). Electrical properties were measured using a PPMS-6000 system (Quantum Design, Japan).

DISCUSSION

Figure 1 shows XRD patterns of the powders prepared from a solution without citric acid (a) and with citric acid (b) by heat treatment in air. In the XRD patterns of the powders prepared from the solution without citric acid and heat-treated at temperatures from 973 to 1073 K, the diffraction peaks were assigned to Y_2O_3 and CuO. We found that precursor phase $Cu_2Y_2O_5$ was generated at above 1173K and that single phase $Cu_2Y_2O_5$ could be obtained at above 1273 K. On the other hand, when the powder prepared from the solution with citric acid was heat-treated at temperatures above 1073 K, we were able to obtain single phase $Cu_2Y_2O_5$. These results show that single phase $Cu_2Y_2O_5$ can be obtained with this method by using the lower temperature range of metal-citric acid complexes and shorter heat treatment time than solid-state reaction. This method using metal-citric acid complexes allows the metal cations to be mixed at the molecular level and stoichiometric compositions to be achieved by chelating the metal ions in solution by citric acid. Further, the organic components of these metal-citric complexes burn and their metal atoms are oxidized at almost the same temperature range. Thus this method enabled precursor phase $Cu_2Y_2O_5$ to be obtained easily and at a low temperature range. Under the above-described heat treatment conditions, however no $CuYO_2$ phase could be obtained from the powders prepared from the solutions with or without citric acid, because in general $CuYO_2$ can be obtained only heat-treating precursor phase $Cu_2Y_2O_5$ in a low partial oxygen pressure atmosphere. The precursor phase powders will decompose to Y_2O_3 and CuO at heat treatment temperatures above 1473 K.

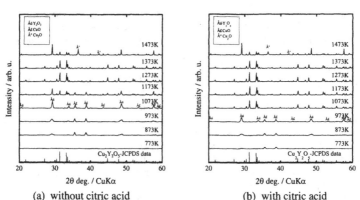

(a) without citric acid   (b) with citric acid

Figure 1 XRD patterns of the powders prepared from a solution without citric acid (a) and without citric acid (b).

Figure 2 shows XRD patterns of the powders prepared from non-precursor powder (a) and precursor $Cu_2Y_2O_5$ powder (b) by heat treatment in an N_2 gas atmosphere. For XRD pattern (a), we investigated the intense peaks assigned to Y_2O_3, Cu_2O and CuO in the low temperature range. Precursor phase $Cu_2Y_2O_5$ could be obtained at above 1073 K and it changed to a $CuYO_2$ phase at slightly higher heat treatment temperature i.e., above 1223 K. On the other hand, single phase $CuYO_2$ without any impurity phases could be obtained from the $Cu_2Y_2O_5$ precursor by heat treatment the latter at above 1223K. The XRD patterns of $CuYO_2$ with 2H (hexagonal)- and 3R (rhombohedral) -polytype delafossite structures reported in the Powder Diffraction File of the International Center for Diffraction Data are shown at the bottom of Fig. 2 (a) and (b) [15,16]. The color of the $CuYO_2$ powder was light brown. In comparing our results with these data, it became apparently that the samples we prepared at temperatures range between 1223 and 1373 K consisted of $CuYO_2$ with 2H and 3R phase. The amount of 2H phase including the sample powder was larger than that of 3R phase.

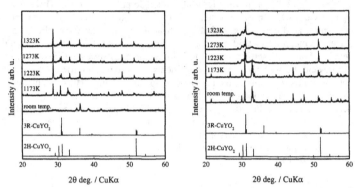

(a) prepared from non-precursor   (b) prepared from precursor $Cu_2Y_2O_5$

Figure 2 XRD patterns of the powders prepared from non-precursor powder (a) and precursor $Cu_2Y_2O_5$ powder (b) by heat treatment in an N_2 gas atmosphere.

The color of the Ca doped $CuYO_2$ powders were light green though that of non-doped sample was light brown. The color of the Ca doped $CuYO_2$ powders became lighter as the amount of Ca for doping was increased. This leads us to believe that the doping changes the valence state of $CuYO_2$. Figure 3 shows XRD patterns of $CuY_{1-x}Ca_xO_2$ ($0 \leq x \leq 0.05$) powders. The diffraction peaks in the patterns were slightly shifted by Ca doping the $CuYO_2$ and then further shifted toward a high diffraction angle as the amount of doping Ca was increased. XRD intensity profiles of the samples seem to change slightly with an increasing amount of doping Ca. We found that the amount of 2H phase increased and that of 3R phase as decreased as amount of doping Ca became higher. These results made it clear that Ca could be substituted for the Y site of $CuYO_2$ without any impurity phases.

Figure 4 shows SEM images of $CuYO_2$ powder (a) and a cross section of the ceramics. The grains were plate-like in shape. We believe that the shape of crystal grains prepared from the solution was affected more by the crystal structure than in other methods. We had considered that the plated-like particles could be easily aligned by the application of uniaxial pressure. However, the grains comprising the ceramics were found to be spherical particles. We believe that plate-like particles changed to spherical shape because their surface free energy decreased with extended heat treatment. We investigated many pores in the SEM image of the cross section of the ceramics. The relative densities of the ceramics were $CuYO_2$: 57.8 %, $CuY_{0.99}Ca_{0.01}O_2$: 66.5% and $CuY_{0.98}Ca_{0.02}O_2$: 66.4%, respectively. The relative densities of the ceramics increased with the addition of Ca. We consider that the reason for this is that Ca doping caused an increasing diffusion coefficient in the interface of grains because of defects on and in the crystal grains.

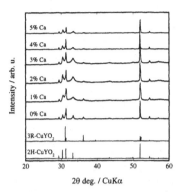

Figure 3 XRD patterns of $CuY_{1-x}Ca_xO_2$ ($0 \leq x \leq 0.05$) powders.

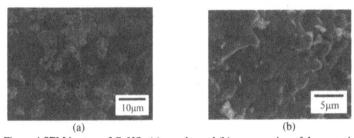

(a)   (b)

Figure 4 SEM images of $CuYO_2$ (a) powder and (b) cross section of the ceramics.

Table shows electrical conductivity of $CuYO_2$, $CuY_{0.99}Ca_{0.01}O_2$ and $CuY_{0.98}Ca_{0.02}O_2$. Though their electrical conductivity levels were very low, they increased with an increasing amount of doping Ca. We consider that the low electrical conductivity was caused by low bulk density or low carrier concentration. Though the sintering temperature we used in our study was somewhat higher than that used in other related studies, we could not obtained highly density ceramics. Usually, the diffusion coefficient in or on the grain increases as the number of crystal structure defects becomes higher. In a report by Cava et al on polycrystalline $CuYO_{2+\delta}$ properties, it was found that $CuYO_{2+\delta}$ doped with calcium showed conductivity as high as 10S/cm after oxygen intercalation. Intercalation of excess O^{2-} ions in the interstitial sites may trap electrons leaving behind empty states in the valence band that act as holes. It has been reported by Nagarajyan et al that the value of x i.e. the percentage of excess oxygen, may be no more than 25% in $CuYO_{2+x}$ polycrystalline powder. On the other hand, in our study case, it is considered that $CuYO_{2+\delta}$ could not be obtained because the Ca doped $CuYO_2$ samples were not annealed in oxygen. Further, we did not investigate the $CuYO_{2+\delta}$ diffraction patterns shown in Fig. 3. Thus, the bulk ceramics did not show highly electrical conductivity and had highly density because we could not obtained oxygen intercalated $CuYO_{2+\delta}$.

Table. Electrical conductivity and relative density of $CuYO_2$, $CuY_{0.99}Ca_{0.01}O_2$ and $CuY_{0.98}Ca_{0.02}O_2$.

Sample	$CuYO_2$	$CuY_{0.99}Ca_{0.01}O_2$	$CuY_{0.98}Ca_{0.02}O_2$
Relative density (%)	57.8	66.5	66.4
Electrical conductivity (S/cm)	-	3.61×10^{-5}	1.09×10^{-4}

CONCLUSIONS

We obtained $Cu_2Y_2O_5$ as a precursor for preparing delafossite $CuYO_2$ could be obtained at low temperature range and short heat treatment time through use of a metal-citric acid complex decomposition method rather than a solid-state reaction method. Delafossite $CuYO_2$ was synthesized by heat treatment above 1223 K in an N_2 gas atmosphere. We were also able to use Ca doping to the valence state of $CuYO_2$ to its Y site. Though the electrical conductivities of our sample materials were very low, they were increased with increasing the amount of doping Ca. It was difficult to obtain dense $CuYO_2$ ceramics by heat treatment temperatures below the decomposition temperature range.

REFERENCES

1. I. Terasaki, Y. Sasago, K. Uchinokura, Phys. Rev. B 56 (1997) R12685-R12687
2. H. Yakabe, K. kikuchi, I. Terasaki, Y. Sasago, K. Uchinokura, Proceedings of the 16th International Conference on Thermoelectrics, (1997) 523-527
3. S. Li, R. Funahashi, I. Matsubara, K. Ueno, S. Sodeoka, H. Yamada, J. Mter. Chem., 9 (1999) 1659-1660
4. R. Manoj, M. Nisha, K. A. Vanaja and M. K. Jayaraj, Bull. Mater. Sci., 31 (2008) 49-53
5. N. Tsuboi, H. Ohara, T. Hoshino, S. Kobayashi, K. Kato and F. Kaneko, J. J. Appl. Phys., 44 (2005) 765-768
6. G. M. Kale and K. T. Jacob, Chem. Mater., 1 (1989) 515-519
7. N. Tsuboi, T. Hoshino, H. Ohara, T. Suzuki, S. Kobayashi, K. Kato and F. Kaneko, J. Phys. Chem. Solid, 66 (2005) 2134-2138
8. M. K. Jayaraj, A. D. Draeseke, J. Tate and A. W. Sleight, Thin Solid Films, 397 (2001) 244-248
9. M. P. Pechini, US Patent (1967) 3330697
10. M. S. G. Baythoun, F. R. Sale, J. Materials Science 17 (1982) 2757-2769
11. L. W. Tai, P. A. Lessing, J.Mater. Res., 7 (1992) 502-510
12. S. Roy, W. Sigmund and F. Aldinger, J. Mater. Res. 14 (1999) 1524-1531
13. Keishi Nishio, Kazuma Takahashi, Yusuke Inaba, Mariko Sakamoto, Tsutomu Iida, Kazuyasu Tokiwa, Yasuo Kogo, Atsuo Yasumori, Tsuneo Watanabe, 23rd International Conference on Thermoelectrics-ITC2004 Proceedings, Released 2005
14. T. Yoshimura, K. Nishio, K. Takahashi, K. Tokiwa, T. Kineri, A. Yasumori, and T. Watanabe, Transaction of the Materials Research Society of Japan, 30 (2005) p519-522
15. JCPDS No.37-929
16. JCPDS No.39-244

Mater. Res. Soc. Symp. Proc. Vol. 1166 © 2009 Materials Research Society   1166-N09-04

Thermoelectric Power Factor of Polycrystalline $La_{0.75}Sr_{0.25}Co_{1-x}Mn_xO_{3-\delta}$ Ceramics

J.E. Rodríguez and J. A. Niño
Department of Physics, Universidad Nacional de Colombia
Thermoelectric Materials Group
Apartado Aéreo 85815, Bogotá, Colombia

ABSTRACT

Thermoelectric properties of polycrystalline $La_{0.75}Sr_{0.25}Co_{1-x}Mn_xO_{3-\delta}$ ($0<x<0.08$) (LSCoO-Mn) compounds have been studied. The samples were grown by solid-state reaction method; their transport properties were studied in the temperature range between 100 and 290K, as a function of temperature and the manganese content. The Seebeck coefficient (S) is positive over the measured temperature range and its magnitude increases with the manganese content up to values close to 160 μV/K. The electrical resistivity (ρ) goes from metallic to semiconducting behavior as the Mn level increases, at room temperature, $\rho(T)$ exhibit values less than 4mΩ-cm. From $S(T)$, $\rho(T)$ and $\kappa(T)$ data, the thermoelectric power factor and the figure of merit were determined. These performance parameters reach maximum values around 18 μW/K^2-cm and 0.2, respectively. The observed behavior in the transport properties become these compounds potential thermoelectric materials, which could be used in thermoelectric applications.

INTRODUCTION

The Seebeck and Peltier effects offer an alternative pathway for solid-state energy conversion. The best thermoelectric materials available near room temperature are bismuth telluride and its alloys. However, these materials are not stable at high temperatures. On the contrary, at low temperature the best thermoelectric materials are monocrystalline Bi-Sb alloys, which are n-type semiconductors. However, their use in practical thermoelectric devices is limited by both their poor mechanical properties and because no suitable p-type material has been found with compatible properties.

The research for efficient thermoelectric materials often involved no conventional semiconductors [1-3]. In this sense, oxide compounds as: $LaCoO_3$, $La_{1-x}Sr_xCoO_3$, $Bi_2Ca_2Co_2O_x$, $NaCo_2O_4$, $Ca_3Co_4O_9$, $Ca_3Co_2O_6$ are promissory candidates to become thermoelectric materials, because of their transport properties and their physical-chemical stability [4-6].

The $La_{1-x}Sr_xCo_{1-x}Mn_xO_{3-\delta}$ compounds are members of perovskite-family, which show a marked asymmetry of the transport properties and a metallic or semiconducting behavior, which depend on the Sr and Mn content and critically on the oxygen stoichiometry [7].

The energy conversion performance of a thermoelectric device is evaluated using the dimensionless figure of merit ZT, which is defined as [1,2]:

$$ZT = \frac{S^2 T}{\rho \kappa} \tag{1}$$

where S is the Seebeck coefficient, ρ the electrical resistivity, κ the total thermal conductivity and T the absolute temperature.

Taking into account the versatility of their transport properties, these LSCoO ceramics can be considered as candidates for thermoelectric applications. In this sense, here, a study of thermoelectric properties of LSCoO-Mn compounds is presented, in which their transport properties were modified through a partial substitution of Co for manganese atoms.

EXPERIMENT

Samples with nominal composition $La_{0.75}Sr_{0.25}Co_{1-x}Mn_xO_{3-\delta}$ ($0 \leq x \leq 0.08$) were prepared by using ceramic routes from high purity powders of La_2O_3, SrO, Co_3O_4 and MnO_2 (Merck 99.99% pure).

Seebeck coefficient data were obtained by differential technique, with an accuracy about 0.5 μV/K. The longitudinal heat flow method was used in the thermal conductivity measurements. To avoid heat loss via residual gases, the $\kappa(T)$ measurements were carried out under vacuum atmosphere (10^{-3} mb). Additionally, the measured thermal conductivity was corrected for radiation losses based on $\sim T^3$ temperature dependence [8,9]. The accuracy of thermal conductivity data was about 0.2 W/K-m.

On the other hand, the electrical resistivity was measured by the standard DC four probe method. These transport properties measurements were carried out in the temperature range between 100 and 290K. In addition, the structural and morphological properties of the samples were studied by powder x-ray diffraction analysis (XRD) and Scanning Electron Microscopy (SEM), respectively.

DISCUSSION

The x-ray diffraction analysis shows a compound with rhombohedral (hexagonal) symmetry having a space group $R\bar{3}c$. No evidence of impurities or secondary phases was found in the spectra (see figure 1). The obtained lattice parameters do not change significantly with the manganese doping, as can be seen in table 1.

Table 1. Lattice parameters of manganese doped LSCoO compounds obtained from powder x-ray diffraction analysis.

Sample	a=b (Å)	c (Å)	γ (deg.)
Mn=0.00	5.396(1)	13.294(4)	120.1
Mn=0.02	5.401(1)	13.297(4)	120.2
Mn=0.03	5.419(1)	13.292(2)	120.2
Mn=0.05	5.397(1)	13.294(5)	120.2
Mn=0.08	5.396(2)	13.294(5)	120.3

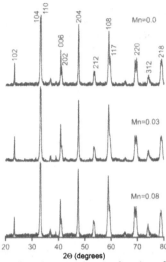

Figure 1. XRD patterns of LSCoO-Mn samples as a function of Mn content.

Figure 2 shows the morphology of LSCO-Mn compounds obtained by scanning electron microscopy. Non-doped samples exhibit the grain structure and the porosity typical of these ceramic compounds. It is clearly seen that with increasing manganese content the grain size decreases from 0.5 μm to 0.44 μm (see table 2), which have effects on the heat flow across the samples.

Figure 2. SEM micrographs of LSCO-Mn compounds grown by solid-state reaction method.

Table 2. Behavior of grain size LSCoO-Mn ceramics as a function of the manganese level.

Sample	Mn=0.00	Mn=0.02	Mn=0.03	Mn=0.05	Mn=0.08
Grain Size (μm)	0.50(1)	0.50(1)	0.48(1)	0.45(1)	0.44(1)

The electrical resistivity decreases with increasing temperature showing a semiconducting-like behavior (see figure 3a); which increases with Mn level. At room temperature, the magnitude of ρ(T) increases with the manganese content from 0.5 mΩ-cm to 3 mΩ-cm.

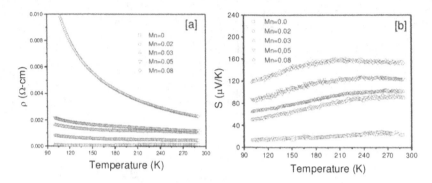

Figure 3. Temperature behavior of electrical resistivity [a] and Seebeck coefficient [b] for LSCoO-Mn ceramics as a function of manganese content.

If the carriers behave as small polarons, the electrical resistivity can be expressed as [10,11]

$$\rho(T) = \frac{1}{pe\mu_p} \approx T \exp\left(\frac{E_a}{k_B T}\right) \tag{2}$$

where μ_p is the movility of holes and E_a is the activation energy.

Plotting $ln(\rho/T)$ as a function of $1/T$, it is possible to calculate the activation energy, which increases with the Mn level, as the semiconducting properties are taking place in the samples (see table 3). This trend agrees with the behavior of Seebeck coefficient of these LSCO-Mn ceramics, which is perhaps related with the decrease of carrier concentration as the manganese level increases.

Table 3. Activation energy E_a as a function of manganese content, obtained from fitting the electrical resistivity experimental data to $\rho(T) \approx T \exp\left(\frac{E_a}{k_B T}\right)$ model.

Sample	Mn=0.00	Mn=0.02	Mn=0.03	Mn=0.05	Mn=0.08
E_a(meV)	7.69(1)	8.26(2)	12.95(1)	17.78(3)	30.01(2)

The behavior of Seebeck coefficient is shown in the figure 3b, S(T) is positive over the measured temperature range, suggesting a hole-type conduction. Its magnitude increases monotonically with the manganese level reaching maximum values close to 160 μV/K, in the temperature range between 170 and 220K.

Across the measured temperature range, S(T) shows two different trends. Between 200 and 290K S(T) is nearly temperature independent, which can be described by Heikes model, for strong correlated systems [12-14]. In the temperature region between 100 and 200 K, the Seebeck coefficient is temperature dependent; this behavior could be ascribed to variable range hopping (VRH) [14,15].

The thermal conductivity is weak temperature dependent, its magnitude decreases with manganese level from 3.5 to 0.8 W/K-m (see figure 4). This trend agrees with the observed behavior in the electrical resistivity and the changes in the grain structure.

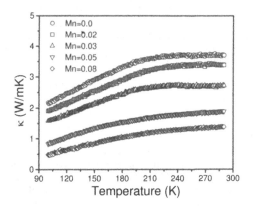

Figure 4. Temperature behavior of thermal conductivity for LSCoO-Mn ceramics.

The thermoelectric power factor PF, is a parameter that determines the electrical properties of a thermoelectric material; this performance parameter is defined as:

$$PF = \frac{S^2}{\rho} \qquad (3)$$

where S and ρ are the Seebeck coefficient and the electrical resistivity, respectively.

Figure 5a shows the temperature behavior of power factor, PF increases with manganese doping up to maximum values close 18 μW/K²-cm, in the temperature range between 200 and 290K for the samples with Mn=0.02.

From S(T), ρ(T) and κ(T) data, the dimensionless figure of merit, ZT was determined (equation 1), the magnitude of ZT increases with the temperature up to reach values close to 0.2 in the temperature range between 250 and 290K, in the samples with Mn=0.08 and 0.05 (see figure 5b). The magnitude and the temperature behavior exhibited by PF and ZT suggest these LSCoO ceramics could be promising thermoelectric material, if they are submitted to proper changes in their chemical composition.

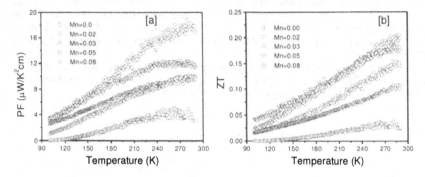

Figure 5. Temperature dependence of thermoelectric power factor [a] and dimensionless figure of merit [b] for LSCO-Mn samples.

CONCLUSIONS

In summary, $La_{0.75}Sr_{0.25}Co_{1-x}Mn_xO_{3-\delta}$ polycrystalline samples were prepared by solid-state reaction method. The transport properties of the samples were modified by changing Co atoms for manganese atoms. The obtained LSCoO-Mn samples exhibit a rhombohedral (hex) structure and high Seebeck coefficient values, which increase with Mn level. The electrical resistivity shows a semiconducting-like behavior and its magnitude is less than 3 mΩ-cm at room temperature. The magnitude of total thermal conductivity decreases with the Mn content, which agrees with the observed behavior of $\rho(T)$ and the grain structure. The studied transport properties yield maximum values for PF and ZT of 18 $\mu W/K^2$-cm and 0.20, respectively, which become this kind of ceramic compounds promising material for thermoelectric applications.

ACKNOWLEDGMENTS

The authors would like to thank to *"La división de Investigación de la Universidad Nacional de Colombia, Sede Bogotá"*, DIB for the financial support.

REFERENCES

1. D. M. Rowe, *CRC handbook of thermoelectrics*, CRC Press, Boca Raton Fl, 1995.
2. G. S. Nolas, J. Sharp and H. J. Goldsmid, *Thermoelectrics, basic principles and new materials developments*, Springer-Verlag, Berlin,2001.
3. G. Mahan, B. Sales and J. Sharp,*Physics Today*,**50**, 42(1997).
4. S. Yamanaka, H. Kobayashi and K. Kurosaki, *J.J. Alloys Comp.***349**,321-324(2003); S. Li, R. Funahashi, I. Matsubara,H. Yamada, K. Ueno and M.Ikebe, *Ceramics International*, **27**, 321-324(2001).
5. E. Sudhakar, J.G. Noudem, S. Hebert and C. Goupil, *J.Phys. D: Appl. Phys.* **38**,3751-3755(2005).
6. L.C. Moreno, D. Cadavid and J. E. Rodriguez, *Microelectronics Journal*, **39**,548-550(2008).
7. J.S. Kim, B.H. Kim, D.C. Kim and Y.W. Park, *Ann.Phys.* (Leipzig) **13**,43-47(2004).

8. A.L. Pope, B.Zawilski and T.M. Tritt, *Cryogenics*,**41**,725-731(2001).

9. *Thermal Conductivity, Theory, Properties and Applications*, Edited by Terry M. Tritt, Springer,New York, 2004.

10. M.A. Señaris-Rodriguez, J.B. Goodedenough, *J. Solid State Chem.* **116**, 224-230 (2001).

11. A.J. Zhou, T.J. Zhu, X.B. Zhao, H.Y. Chen and E. Muller, *Alloys and Compounds*, **449**,108-108(2008).

12. J.F. Kwak, *Phys. Rev. B*, **13**, 652-657(1976).

13. P.M. Chaikim and G. Beni, *Phys. Rev. B*, **13**,647-651(1976).

14. N.F. Mott and E.A. Davis, *Electronic Processes in non-crystalline materials*, Claredon Press, Oxford, 1979.

15. N.V. Lien and D.D. Toi, *Phys. Lett. A*, **261**, 108-113(1999).

Mater. Res. Soc. Symp. Proc. Vol. 1166 © 2009 Materials Research Society   1166-N09-07

Melt-Textured Growth of Grain Aligned Bulk Oxide Thermoelectrics

Venkat Selvamanickam and Bo Zhang
Department of Mechanical Engineering
4800 Calhoun Road
University of Houston
Houston, Texas 77204, U.S.A.

ABSTRACT

A new process, based on melt processing has been investigated with oxide thermoelectrics to achieve long-range grain alignment for low resistivity as well as to embed secondary phase precipitates and associated crystal defects for low thermal conductivity. Melt-processing of $Bi_2Sr_2Co_2O_x$ and $CaMnO_3$ has been studied. A high degree of (00l) grain alignment has been achieved by melt processing. Good values of Seebeck coefficient of nearly -250 $\mu V/K$ were measured in the melt-processed $CaMnO_3$. Secondary phases of $Ca_4Mn_3O_{10}$ and $CaMn_2O_4$ are found to be trapped between the aligned grains which led to high electrical resistivity values and limited the figure of merit in our initial samples.

INTRODUCTION

The Figure of merit (ZT) values of thermoelectrics have typically hovered around one in bulk materials. Bulk materials are needed for large-scale generation of electric power from waste heat and in spite of significant progress [1], achieving ZT values higher than three is challenging. The challenge stems from conflicting performance requirements in thermoelectric materials in general, to achieve high electrical conductivity, but low thermal conductivity while maintaining a high Seebeck coefficient. Additional challenges have been faced with the deteriorating properties at high temperatures, instability of the structures at operating conditions, use of toxic materials such as Pb, Te, Se etc, high cost of Te, poor oxidation resistance and lack of mechanical robustness.

Oxides are well suited for high-temperature operation since their ZT increases even at temperatures of 1000 K. Also, oxides exhibit excellent stability and oxidation resistance at high temperatures, that are key requirements for thermoelectric power generation from waste heat. Furthermore, thermoelectric oxides do not include Pb, Te, Se etc and can be processed in air instead of sealed quartz tube conditions. But, the ZT values of oxides have been low so far. So, we have recently begun investigation of a novel process technique based on melt-textured growth that deviates from the typical ceramic processing methods used so far.

Since oxide thermoelectrics are highly anisotropic, electrical conductivity along the basal plane is about two orders of magnitude higher than that perpendicular to the plane [2, 3]. In another study, the resistivity along the basal plane of Na_xCoO_{2-d} single crystals was found to be an order of magnitude lower than that in polycrystalline bulk ceramic samples [4]. However, the thermal conductivity of single crystals has not been reduced due to the lack of phonon scattering defects such as grain boundaries, second-phase precipitates, and dislocations. The thermal conductivity of single crystal $Na_xCoO_{2-\delta}$ is 2.4 times higher than that of $Na_xCoO_{2-\delta}$ polycrystals [4]. If its thermal conductivity can be reduced to the level of polycrystalline material, then the ZT at 800 K would be nearly three. Bulk sintered materials with fine grains and nano-scale secondary phases have exhibited lower thermal conductivity, but have displayed lower electrical conductivity because of random grain orientation and the prevalence of grain boundaries. The main hypothesis of our approach is that a microstructure that can be created with an excellent grain alignment of anisotropic thermoelectric oxides along with fine-scale defects such as precipitates, dislocations, and stacking faults will meet the conflicting performance requirements in thermoelectrics and lead to the elusive high figure of merit at high temperatures.

EXPERIMENT

A top-seeded melt-texturing technique, previously developed for oxide superconductors, is the process we are exploring to introduce long-range grain alignment [5, 6]. This technique is schematically shown in Figure 1, where a sintered thermoelectric oxide is subjected to partial melting with a short excursion in temperature above its melting point. A single crystalline or a melt-textured seed of a material with a melting point above this temperature, and with a same crystallographic structure, and lattice parameters close to that of the thermoelectric oxide is used. Nucleation is initiated from the top of the partially-melted thermoelectric sample as it is cooled down to the vicinity of its solidification temperature. By slow cooling through the solidification temperature or holding at an undercooled temperature, growth of crystallites with grains aligned along the basal plane proceeds laterally and vertically with an epitaxial relationship from the seed. Essentially, the crystallographic orientation of the seed material is reproduced from a size of a few millimeters to about 10 cm in the bulk oxide. Since the basal planes of the melt-textured material could be aligned over the entire bulk, a high electrical conductivity is expected, which is important for high ZT. Melt-textured growth, especially of compounds that undergo peritectic solidification, also provides pathways to create nano-scale defects such as precipitates, sub grain boundaries, dislocations, and stacking faults which could all lower thermal conductivity through phonon scattering in thermoelectrics just like it dramatically improved critical current performance in bulk superconductors through fluxon trapping [7].

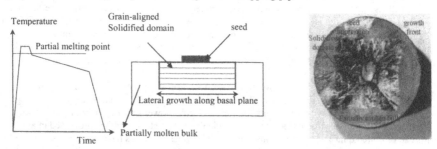

Figure 1. *Melt-textured growth process being explored to fabricate grain-aligned bulk thermoelectric oxides. Also included in a top view of a melt-textured superconductor showing a planar growth front of a single solidified domain.*

Two oxide thermoelectric materials were investigated in this study : $Bi_2Sr_2Co_2O_x$ and $CaMnO_3$. $CaMnO_3$ is a perovskite like $YBa_2Cu_3O_x$ and has exhibited a good Seebeck coefficient value of -150 $\mu V/K$ at 1000 K, but its electrical and thermal conductivity values need to be improved [8]. $CaMnO_3$ is also of interest because of mixed valence of Mn which allows doping to modify its carrier concentration and mobility [9]. Recently, it has been shown that its ZT can be improved by Nb substitution in a *sintered ceramic, polycrystalline form* [10].

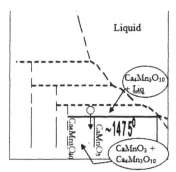

Figure 2 *Phase diagram of CaMnO₃ from reference 11.*

A portion of the phase diagram of CaO – Mn₂O₃ system [11] is shown in Figure 2. As shown in the figure, CaMnO₃ undergoes incongruent peritectic melting at 1475°C to Ca₄Mn₃O₁₀ and liquid, which is a characteristic that is favored for this work for creating a microstructure of precipitates of Ca₄Mn₃O₁₀ in a grain-aligned matrix of CaMnO₃. After partial melting, either a slow cooling temperature profile through the peritectic solidification temperature or an isothermal hold at an undercooled condition is used to form grain-aligned CaMnO₃.

Powders for this study were prepared by solid-state synthesis routes. In the case of $Bi_2Sr_2Co_2O_x$, the powders were calcined for 20 hours each at 800°C and 820°C and then pressed into pellets followed by sintering at 840°C for 20 hours. Melt-textured growth was conducted over a temperature range of 950 to 875°C with slow cooling rates of 1 to 2°C/hour. In addition, we also used spark plasma sintering (SPS) of the powders at 820°C for 5 minutes. In the case of CaMnO₃, 2 mol% Nb was added to the powders which were calcined for 15 hours each at 1000°C and 1300°C and then pressed into pellets followed by sintering at 1400°C for 12 hours. Melt-textured growth was conducted over a temperature range of 1560 to 1450°C with slow cooling rates of 1 to 2°C/hour.

RESULTS AND DISCUSSION

X-ray Diffraction (XRD) data and Scanning Electron Microscopy (SEM) conducted on a SPS sample of $Bi_2Sr_2Co_2O_x$ are shown in Figure 3. The phases present in the sample were essentially all attributed to $Bi_2Sr_2Co_2O_x$ which shows random orientation as expected. The microstructure was found to consist of small grains about 100 nm in dimension which was anticipated from the short duration of the SPS process. Thermoelectric property measurements

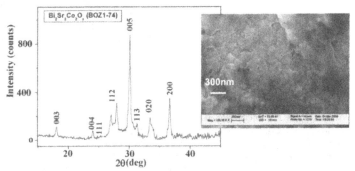

Figure 3. *XRD and SEM data from a spark plasma sintered $Bi_2Sr_2Co_2O_x$. Random orientation of small grains is seen.*

obtained from the SPS sample are summarized in Figure 4. As shown in the Figure, the Seebeck coefficient was found to increase beyond 800 K up to a value of 210 μV/K at 950 K and the thermal conductivity was found to increase steadily with temperature reaching 1.27 W/m/K at 950 K. The resistivity was seen to increase at the higher temperatures to a value of about 19 mΩcm at 950 K. In spite of the good Seebeck coefficient, the ZT value was only 0.18 at 950 K.

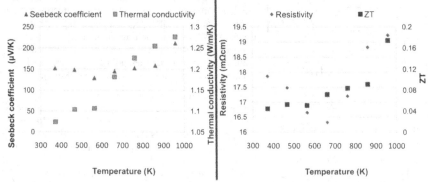

Figure 4. *Thermoelectric property measurements from a spark plasma sintered Bi$_2$Sr$_2$Co$_2$O$_x$.*

Our initial melt-texturing experiments with *thermoelectric* Bi$_2$Sr$_2$Co$_2$O$_x$ show grain-aligned microstructures and good (00l) texture from X-ray Diffraction measurements (see Figure 5). A remarkable difference is seen from the data from sintered sample shown in Figure 3.

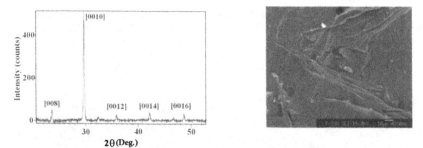

Figure 5. *Texture and microstructure data obtained from initial experiments of melt-textured thermoelectric Bi$_2$Sr$_2$Co$_2$O$_x$. Grain-aligned microstructure and a (00l) texture can be seen.*

Melt-texturing of 2%Nb:CaMnO$_3$ was found to be more complicated due to significant differences along the thickness of the sample which is evident from the XRD patterns shown in Figure 6. A preferred single (002) orientation of CaMnO$_3$ is observed at the bottom of the melt-processed sample. Remnant Ca$_4$Mn$_3$O$_{10}$ and CaMn$_2$O$_4$ are present indicating the sluggish nature of peritectic reaction in the formation of CaMnO$_3$. No preferred orientation of CaMnO$_3$ is observed at the top surface. In fact, a substantial amount of secondary phases are observed. The microstructure of the top surface (Fig. 7) shows needle-shaped secondary phases embedded in a dense matrix with no particular grain morphology. In the bottom surface, several large crystallites of CaMnO$_3$ are seen but secondary phases are found to be trapped between the grains.

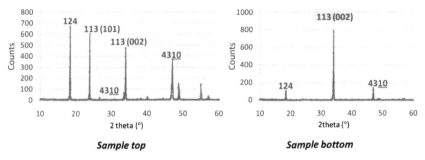

Sample top

Sample bottom

Figure 6. *XRD patterns obtained from melt-processed Nb:CaMnO₃ sample*

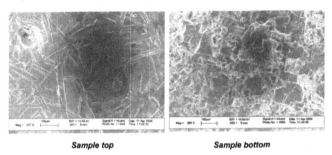

Sample top

Sample bottom

Figure 7. *Surface microstructures of melt-processed Nb:CaMnO₃ sample*

Thermoelectric property measurements were conducted on three melt processed and one sintered 2%Nb:CaMnO₃ samples and the results are shown in Figure 8. The primary difference among the three melt-processed samples were the slow cooling schemes (sample 1 : 2°C/h from 1550°C to 1500°C, sample 2 : 10°C/h from 1550°C to 1525°C and 2°C/h from 1525°C to 1500°C, and sample 3 : Longer (30 minute) hold at 1560°C and 3°C/h from 1550°C to 1500°C). As shown in Figure 8, all melt-processed samples had higher resistivity and lower thermal conductivity than

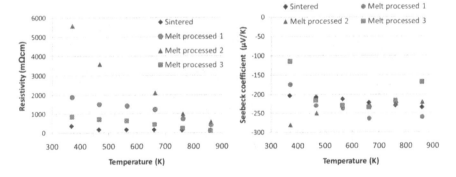

215

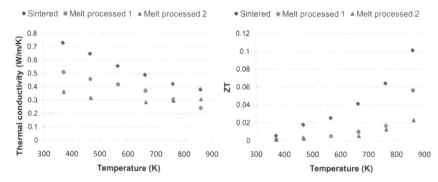

Figure 8. *Thermoelectric property data from melt-processed and sintered 2%Nb:CaMnO₃*

the sintered sample most likely because of the trapped secondary phases between the large CaMnO$_3$ grains. Good values of Seebeck coefficient (~ -200 to – 250 μV/K) were measured in all samples. Due to the higher resistivity, the ZT values of the melt-processed samples were inferior to that of the sintered sample. Hence, in order to benefit from the grain alignment in the melt-processed samples, the secondary phases between the aligned grains have to be eliminated. We are working on modified compositions and process parameters to eliminate the secondary phases and refine them to embed them as final-scale precipitates within the matrix.

ACKNOWLEDGMENTS

We are very grateful to Prof. Terry Tritt of Clemson University for providing us access to facilities for thermoelectric property measurements as well as to Dr. V. Ponnambalam, Mr. T. Colgate, and Mr. W. Xie for providing support with the measurements.

REFERENCES

1. G. S. Nolas, J. Poon,and M. Kanatzidis, *Mat. Res. Soc. Bull,* **31,** 199 (2006)
2. T. Terasaki, Y. Sasago, K. Uchinokura, *Phys. Rev. B.* **56,** R1268 (1997)
3. R. Funahashi, I. Matsubara, H. Ikuta, T. Takeuchi, U. Mizutani, and S. Sodeoka, *Jpn. J. Appl. Phys.* **39,** L1127 (2000)
4. K. Fujita, T. Mochida, K. Nakamura, *Jpn. J. Appl. Phys.* **40,** 4644 (2001)
5. K. Salama, V. Selvamanickam, L. Gao, and K. Sun, *Appl. Phys. Lett.* **54,** 2352 (1989)
6. V. Selvamanickam, D. Kirchoff, C. E. Oberly, K. Salama, Y. Zhang, and S. Salib, in *High Temperature Superconductors : Synthesis, Processing, and Applications,* ed. U. Balachandran and P. J. McGinn, TMS, Warrendale, 1997, p. 117.
7. M. Murakami, M. Morita, K. Doi, and K. Miyamoto, *Jpn. J. Appl. Phys.* **28,** 1189 (1989)
8. D. Flahaut, T. Mihara, R. Funahashi, *J. Appl. Phys.* **100,** 084911 (2006)
9. B. Raveau, Y. Zhao, C. Martin, M. Hervieu, A. Maignan, *J. Sol. State Chem.* **149,** 203 (2000)
10. L. Bocher, M. H. Aguirre, D. Logvinovich, A. Shkabko, R. Robert, M. Trottman, and A. Weidenkaff, *Inorg. Chem,* **47,** 8077 (2008)
11. H. S. Horowitz and J. M. Longo, *Mater. Res. Bull.* **13,** 1359-1369 (1978).

Mater. Res. Soc. Symp. Proc. Vol. 1166 © 2009 Materials Research Society   1166-N03-05

Thermoelectric Properties of Nb-Doped SrTiO₃ / TiO₂ Multiphase Composite

Kiyoshi Fuda[1] Kenji Murakami[1], Tomoyoshi Shoji[1] and Shigeaki Sugiyama[2]

[1]Department of Applied Chemistry for Environments, Akita Univ., Akita 010-8502, JAPAN,
[2]Akita pref. Ind. Tech. Center, Akita, 010-1623, JAPAN
fudak@ipc.akita-u.ac.jp

ABSTRACT

In this study, we fabricated and examined a series of multiphase type composites constructed of Nb-doped SrTiO₃ / TiO₂ fine particles. The composition of the composites and the sintering temperatures were selected in a two-phase region where a perovskite SrTiO₃ and a rutile TiO₂ phases coexist. The composites obtained here were found to commonly have a mosaic type texture constructed of TiO₂ and SrTiO₃ fine particles with a typical size of about 500 nm. In some samples we also found additive phases such as $Sr_6Ti_7Nb_9O_{42}$. The thermal conductivity values measured for the most samples with different contents are ranged between 2 and 5 Wm⁻¹K⁻¹. The values are apparently lower than the value for single crystal SrTiO₃ samples presented in literature. A sample with rather low relative density of about 80% showed a quite low thermal conductivity, about 1 Wm⁻¹K⁻¹. Taking account the other TE data, e.g. Seebeck coefficient and electrical conductivity, we calculated dimensionless figure of merit, ZT, to be at maximum 0.24 at 600°C.

INTRODUCTION

Oxide thermoelectric (TE) materials have recently drawn a special interest due to thermal stability as well as environmental friendliness. Layered cobaltites such as $NaCo_2O_4$ have been studied as a promising materials group having high performance of p-type TE property. However, it seems that no satisfactory TE property has been found in n-type oxide bulk materials even though Al-doped ZnO and La-doped SrTiO₃ have high thermoelectric (TE) responses [1,2]. Most recently, Ohta et al. demonstrated that super-lattice materials stacking Nb-doped SrTiO₃ and non-doped SrTiO₃ layers could give very large Seebeck coefficients due to two-dimensionally confined electrons [3]. This finding is quite suggestive for materials design of bulk thermoelectric materials. If we construct a bulk oxide with high density of interface between SrTiO₃ and another phase, the composite oxide could give a high TE property. In addition, the interface between different phases would also cause a high thermal resistivity due to effective phonon scattering at such interface. The purpose of this study is to show a possibility of synthesis of n-type TE oxide bulk materials having low thermal conductivity and excellent TE properties as well. For this purpose, we fabricated and examined a series of composites containing two or more oxide phases such as TiO₂ and Nb-doped SrTiO₃.

EXPERIMENTAL

The composite specimens were prepared by two successive steps: (1) preparation of precursor oxide by wet processes; and (2) sintering by using a spark plasma sintering (SPS) apparatus. The preparation of the precursor oxide was started with a hydrolysis of Ti-tetra-isopropoxide in the presence of $NbCl_5$ in proper proportion. The obtained mixed niobium and titanium oxide gel was suspended in an aqueous solution containing Sr ions in a proper proportion, and further hydrolyzed by addition of $(NH_4)_2CO_3$ to form $SrCO_3$ on the titanium/niobium oxide surface. The resultant powder was separated by filtration, washed with deionized water repeatedly, and dried in air at 70 °C. This powder was further calcined at 1000 °C for 1 hour, giving mixed oxide powder precursors.

The specimens for the thermoelectric and the thermal conductivity measurements were prepared by the SPS sintering at 1350 °C under 50 MPa for 5 min into pellets with different diameters of 10 and 20 mm. The former pellet was used for the thermal conductivity measurement, and the latter pellet was further cut in a shape appropriate for the TE-measurement. In the same way, we prepared 7 samples with different compositions of Sr:Ti:Nb, as shown in Table I.

Table I. Molar ratios of cations in the samples used in the present study and their density

Sample name	Nominal metallic composition Sr : Ti : Nb	Measured metallic composition Sr : Ti : Nb	Measured Density /g cm^{-3}
Sr20 Nb20	0.2 : 0.8 : 0.2	0.11 : 0.79 : 0.21	4.38
Sr25 Nb20	0.25 : 0.8 : 0.2	0.23 : 0.79 : 0.21	4.43
Sr30 Nb10	0.3 : 0.9 : 0.1	0.28 : 0.84 : 0.16	4.37
Sr30 Nb20	0.3 : 0.8 : 0.2	0.30 : 0.76 : 0.24	4.53
Sr30 Nb30	0.3 : 0.7 : 0.3	0.27 : 0.66 : 0.34	4.59
Sr50 Nb20	0.5 : 0.8 : 0.2	0.42 : 0.76 : 0.24	4.76
Sr70 Nb20	0.7 : 0.8 : 0.2	0.63 : 0.77 : 0.23	4.76

The microscopic structure was observed by using scanning electron microscope (SEM; HITACHI S-4500 model) attached with an energy dispersive x-ray spectroscopy. The elemental analysis was carried out with an EPMA measurement. The electrical conductivities and the Seebeck coefficients were measured simultaneously using an ULVAC ZEM-1 instrument in helium atmosphere. The thermal diffusivities were measured by a laser flash method in vacuum. The thermal capacity was estimated numerically by using data for TiO_2 and $SrTiO_3$ phases presented in literature. [4] By using these values, we calculated the thermal conductivity values. The uncertainty of the measurement were estimated as follows: thermal diffusivity: ± 5%; electrical resistivity: ± 10%; Seebeck coefficient: ± 7%.

RESULTS AND DISCUSSION

Phase study and micro texture of the samples
The typical XRD patterns of the sintered sample are presented in Fig. 1. The composites

obtained here were found to contain TiO_2 and $SrTiO_3$ phases commonly. In Sr30Nb30 and Sr20Nb20 samples, an additional phase, $Sr_6Ti_7Nb_9O_{42}$, was also found. This phase has hexagonal structure with lattice constant a = 8.9991 Å and c = 11.5118 Å [5]. In a careful inspection, this phase can be found in the other samples, Sr30Nb20 and Sr25Nb20.

SEM photographs showed that the composites have a mosaic type texture constructed of fine crystal particles with a typical size of 0.5 – 1 µm as shown in Fig. 2. From EPMA spot analysis, we distinguished two different parts in the samples: the light gray part and the dark gray part. The former part is rich in heavy metals, i.e. Sr and Nb, corresponding to $SrTiO_3$ and $Sr_6Ti_7Nb_9O_{42}$ phases. But these two phases could not be distinguished only by this EPMA analysis. As the Sr contents in the samples increased, the fraction of the light gray part portion in the SEM image increased. The latter dark gray part was attributed to TiO_2 (rutile) phase. In both these two different parts, more or less Nb content was found. The lattice constants of $SrTiO_3$ and TiO_2 phases were found to increase about 1 % in length for all samples. This finding suggests that niobium ions are dissolved into both of $SrTiO_3$ and TiO_2 phases in the composites. In addition to these two parts, we also found black spots in the SEM images. They may be assigned to voids.

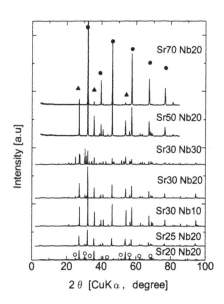

Fig. 1 XRD patterns for Nb-doped TiO_2/$SrTiO_3$ multiphase composites. The keys presented in the figure stands for three phases: •: $SrTiO_3$; ▲: TiO_2 (rutile); ○: $Sr_6Ti_7Nb_9O_{42}$.

219

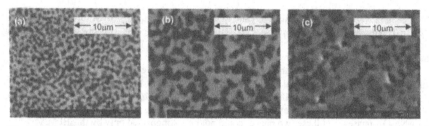

Fig. 2. SEM images of Nb-doped $TiO_2/SrTiO_3$ multiphase composites: (a) Sr25Nb20, (b)Sr30Nb20, (c)Sr50Nb20.

Thermoelectric properties

The TE properties, the Seebeck coefficient, S, and the electrical conductivity, σ, as well as the power factor for Nb-doped $TiO_2/SrTiO_3$ multiphase composites are summarized in Fig.3. The Seebeck coefficients for these samples were all negative values ranging from -150 to -400 μVK^{-1}, suggesting that the all samples are n-type conductor. This poor temperature dependence is in contrast to those for the La-doped $SrTiO_3$ reported by Muta et al [6]. The origin of this small temperature dependence of Seebeck coefficient for our samples is not clear so far, but seems to associated with the thermodynamics of the electron system under coexistence of two or more different phases.

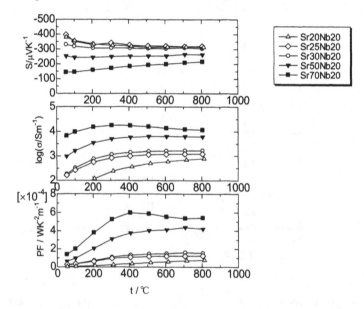

Fig. 3. Thermoelectric properties of Nb-doped $TiO_2/SrTiO_3$ multiphase composites.

Thermal conductivity

The thermal conductivities for the three samples with nominal Sr contents of 0.3 are summarized in Fig. 3(a). In this figure, those for single crystal $SrTiO_3$ samples presented in literature [6] are also presented for comparison. The data for the other four composite samples are not shown, but they were in the range between 3 and 5 $Wm^{-1}K^{-1}$. The values are apparently lower than the value for single crystals. As the Nb content increases, the thermal conductivity increases. It was also found that there is a correlation between the thermal conductivity and the relative density. in this figure, a data for the sample with lower density of 0.843 is also presented. This sample was unintentionally produced. The composition of this sample was Sr : Ti : Nb = 0.3:0.8:0.2 (nominal). Including this one, the data shows a good linear relationship. This result suggests that the porosity plays a important role for reduction of thermal conductivity in this heterogeneous system. Muta et al showed that the thermal conductivity varies depending on the atomic mass of rare earth dopant as well as their ionic radius. In our case, the result presented in Figre 4 (b) suggests that grain boundaries rather than impurities play a predominant role in determining the thermal conductivity.

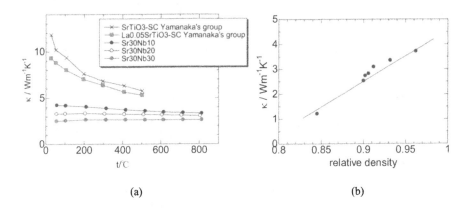

(a)   (b)

Figs. 4 (a):Thermal conductivity of Nb-doped $TiO_2/SrTiO_3$ multiphase composites. For comparison, data for $SrTiO_3$ single crystal samples presented in literature [7] are also presented. (b):Correlation between thermal conductivity at 800 °C and relative density of the composites.

Figure of merit

Taking account the other TE data, e.g. Seebeck coefficient and electrical conductivity, we calculated the dimensionless figure of merit, ZT, as is shown in Fig. 5. Form the figure, Sr70Nb20 sample showed the best performance among the 7 samples. The Sr30Nb20 with low relative density of 0.843 mentioned above gave ZT = 0.24 at 600 °C. The ZT values obtained in this study are not yet satisfactory for practical use. But the reduction in thermal conductivity demonstrated here would give a new possibility to gain oxide thermoelectric materials with higher performance.

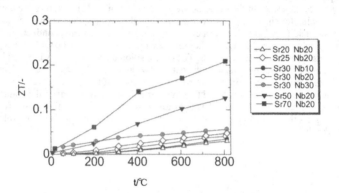

Fig. 5. Dimensionless figure of merits for Nb-doped $TiO_2/SrTiO_3$ multiphase composites.

ACKNOWLEDGMENTS

Financial support from Japan Science and Technology Agency (CREST program) is gratefully acknowledged.

REFERENCES

1. M. Ohtaki, T. Tsubota, K. Eguchi,H. Arai, J. Appl. Phys. **79**, 1816 (1996)
2. T. Okuda, K. Nakanishi, S. Miyasaka, Y. Tokura, Phys. Rev. **B63** 113104 (2001)
3. H.Ohta, S-W. Kim, Y. Mune, T.Mizoguchi, K. Nomura, S. Ohta, T.Nomura, Y. Nakanishi, Y. Ikuhara, M. Hirano, H. Hosono, K. Koumoto, Nature Mater., **6**, 129 (2007)
4. Barin I: "Thermochemical Data of Pure Substances", VCH Verlags Gesellschaft, Weinheim, (1993)
5. H. Zhang, L. Fang, J. Xie, B. Wu, R. Yuan, J. Wuhan Univ. Technol. Mater. Sci. Ed, 15, 59 (2000); JCPDS, 48-859.
6. H. Muta, K. Kurosaki, S. Yamanaka, J. Alloy Comp., **350**, 292 (2003)
7. H. Muta, K. Kurosaki, S. Yamanaka, J. Alloy Comp., **392**, 306 (2005)

Mater. Res. Soc. Symp. Proc. Vol. 1166 © 2009 Materials Research Society   1166-N03-23

Development of Al$_2$O$_3$-ZnO/Ca$_3$Co$_4$O$_9$ Module for Thermoelectric Power Generation

Paolo Mele[1,2], Kaname Matsumoto[1,2], Takeshi Azuma[1], Keita Kamesawa[1], Saburo Tanaka[1], Jun-ichiro Kurosaki[1] and Koji Miyazaki[1,2]

[1]Department of Materials Science and Engineering, Kyushu Institute of Technology (KIT), 1-1, Sensui-cho, Tobata-ku, Kitakyushu 804-8550, Japan
[2]Fukuoka Industry, Science and Technology Foundation (IST), System LSI Division, 3-8-33, Momochihama, Sawara-ku, Fukuoka 814-0001, Japan

ABSTRACT

Pure and Al$_2$O$_3$ (2%, 5%, 8%) doped sintered ZnO (*n*-type) and pure sintered Ca$_3$Co$_4$O$_9$ (*p*-type) pellets were prepared by conventional solid state synthesis starting from the oxides. The sintered pellets were cut by a diamond saw in a pillar shape (15 mm × 5 mm × 5 mm) for physical properties measurements. The best doped sample was 2 % Al$_2$O$_3$ ZnO showing Seebeck coefficient S = -180 mV/K and electrical conductivity σ = 8 S/cm at 400°C, while thermal conductivity κ = 1.8 W/m×K at 600°C. Typical values for Ca$_3$Co$_4$O$_9$ were S = 82.5 mV/K and σ = 125 S/cm at 800°C, while κ = 1.01 W/m × K at 600°C.

Several modules fabricated by elements cut from sintered pellets were tested and the best performance was obtained in the module formed by six 2 % Al$_2$O$_3$-ZnO/ Ca$_3$Co$_4$O$_9$ couples, that generated an output power P = 3.7 × 10^{-5} W at 500°C (when ΔT = 260°C).

INTRODUCTION

These days, many difficult problems appear worldwide: energy problems, environmental problems, and so on. In order to overcome these difficulties, development of new industrial technologies is necessary. These new technologies will play very important roles in many kinds of fields, helping address present and future sustainable energy needs. New materials will play an important role in the current challenge to develop alternative energy technologies to reduce our dependence on fossil fuels and reduce greenhouse gas emissions [1]. An important class of materials is thermoelectrics, that can convert waste heat into electrical energy. One of the most fascinating challenges of material science is to develop new high-performance and low-cost thermoelectric materials for practical applications [2,3]. Then the science and technology of thermoelectric materials is expected to be fundamental in the future [1-3].

The performance of thermoelectric materials is expressed by the thermoelectric figure of merit $ZT = (\sigma S^2 T)/\kappa$, where σ, S, T and κ are the electrical conductivity, Seebeck coefficient, absolute temperature and thermal conductivity, respectively. According to the expression, larger values of σ and S and smaller values of κ are essential for material with higher figure of merit. Recently, extremely high values of ZT were obtained in multilayered nanostructured Bi$_2$Te$_3$/Sb$_2$Te$_3$ thin films [4], with ZT = 2.4 at 300K and in nanostructured BiSbTe bulk alloys with ZT = 1.4 at 100°C [5]. Despite to their outstanding ZT values, these materials contain rare elements [6], are widely unstable at high temperatures (for example, Bi$_2$Te$_3$ decomposes at 857 K [7] and its maximum operating temperature is just 400 K [2]) and require high-cost-processing. On the other hand, looking to practical utilization of thermoelectric materials, *e.g.* the development of modules for thermoelectric power generation, bulk metallic oxides have been recognized as good candidates since they are thermally and electrically stable and low-cost.

In this work, we describe our approach for the development and testing of a module for thermoelectric power conversion using Al_2O_3 doped ZnO [8] and $Ca_3Co_4O_9$ [9] sintered bulk oxides as n-type and p-type materials, respectively.

EXPERIMENTAL

Pure and Al_2O_3 doped sintered ZnO (n-type) and pure sintered $Ca_3Co_4O_9$ (p-type) pellets were prepared by conventional solid state synthesis starting from the oxides. In the case of n-type, ZnO and Al_2O_3 nanometric powders were weighted in proper stoichiometric amounts to obtain 0 mol%, 2 mol%, 5 mol% and 8 mol% Al_2O_3 doped ZnO. The reagents were mixed by means of ZrO_2 balls in ethanol for 6h, then dried at 80 °C for 24h. The dried mixtures were subjected to a preliminary sintering treatment at 1000 °C for 8h, then remixed again in the same conditions used at the beginning. Finally, 25 mm ϕ, 2 mm thick pellets and 10 mm ϕ, 2 mm thick pellets were isostatically pressed and sintered at 1400 °C for 8h. In the case of p-type, CaO and Co_3O_4 were weighted in proper stoichiometric amounts to obtain $Ca_3Co_4O_9$. The synthesis procedure was the same used for ZnO+Al_2O_3 pellets, however the temperatures were 600 °C for the preliminar sintering and 960 °C for the final sintering.

The 25 mm ϕ sintered pellets were cut by a diamond saw in a pillar shape (20 mm×2 mm×2 mm) for following experimental checks. Crystal phase was determined by XRD. Seebeck coefficient (S) and electrical conductivity (σ) were measured in the range 100 °C – 800 °C ($\pm$ 5 °C) by means of ULVAC ZEM-1 apparatus while thermal conductivity (κ) was checked by ULVAC TC9000-H facility on the 10 mm ϕ pellets.

DISCUSSION

Characterisation of samples

Figure 1 shows typical polished cross-sectional SEM image of ZnO+ 2%Al_2O_3 sample. The presence of dark inclusions, which size is about 2~3 μm and fractional area is 4%, is clearly visible. The EDS elemental mapping allows to say that these inclusions are made by $ZnAl_2O_4$ spinel phase. The morphology of samples doped with higher content of Al_2O_3 (5% and 8%) is similar to the morphology of 2% Al_2O_3 sample, and the amount of spurious $ZnAl_2O_4$ phase, which is reported to have low conductivity, [10] increases by increasing the Al_2O_3 content in the samples as expected.

The θ–2θ X-ray diffraction patterns are consistent with SEM analysis since they have ZnO peaks and $ZnAl_2O_4$ peaks while Al_2O_3 peaks were not found. Figure 2 reports the diffraction pattern of ZnO + 2% Al_2O_3 sample.

224

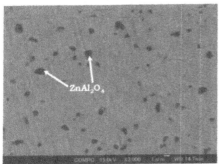

Figure 1. SEM cross-sectional image of ZnO+ 2%Al$_2$O$_3$ sample.

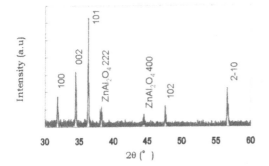

Figure 2. XRD pattern of ZnO+ 2%Al$_2$O$_3$ sample.

Results of the Seebeck coefficient (S) of the sintered samples are shown in Fig. 3 as a function of sample temperature. The observed S was negative for ZnO doped samples, indicating n-type conductivity, and positive for Ca$_3$Co$_4$O$_9$ sample indicating p-type conductivity. Compared with the undoped ZnO sample, 5% Al$_2$O$_3$ doped sample presents a maximum of S while the other samples have smaller S. The order of magnitude of S is quite similar to typical values reported in literature for ZnO samples and Ca$_3$Co$_4$O$_9$ samples [8, 9].

Results of the electrical conductivity (σ) of the sintered samples are shown in Fig. 4 as a function of sample temperature. Except for Ca$_3$Co$_4$O$_9$ sample, the value of σ is quite small if compared with previous reports [8, 9].

Results of the thermal conductivity (κ) of the sintered samples are shown in Fig. 5 as a function of sample temperature. In this case, we succeeded to have quite small values of κ especially in the 2% Al$_2$O$_3$ doped sample which performance approaches Ca$_3$Co$_4$O$_9$.

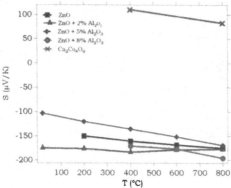

Figure 3. Temperature dependence of the Seebeck coefficient S of sintered samples

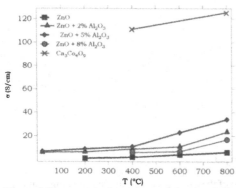

Figure 4. Temperature dependence of the electrical conductivity σ of sintered samples

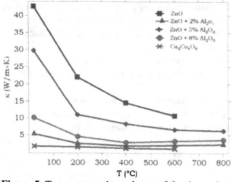

Figure 5. Temperature dependence of the thermal conductivity κ of sintered samples

226

Fabrication and checking of module

   Both n-type and p-type pellets were cut in small pieces (10 mm×1 mm×1 mm) that were electrically connected in series to evaluate the power conversion capability of a module. Couples of n-type and p-type elements were stuck on Ag electrodes (10 mm×10 mm) by silver paint. The n-p couples were then connected by Ag plates and heated from the bottom on a hot plate (T_{MAX} up to 600 °C) while monitoring voltage and temperature. Several modules were fabricated increasing the number of connected couples up to six. A fin was placed on the top of the elements to keep a ΔT between the elements ends.
Photographs of the module together with a cartoon of the module arrangement are reported in Figure 6.

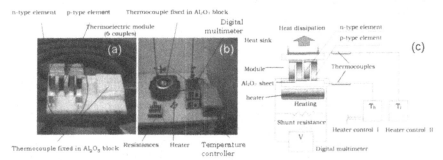

Figure 6. (a) Close view of the module constituted by 6 couples of 2%Al_2O_3- ZnO/ $Ca_3Co_4O_9$. (b) Picture and (c) Schematic representation of the arrangement for the module check

The best performance was obtained in the module formed by 2 mol% Al_2O_3-ZnO /$Ca_3Co_4O_9$ couples, that generated a voltage V = 300 mV at 500°C (when ΔT = 260°C). The output power was P=3.7 × 10^{-5} W and the dimensionless figure of merit of module was ZT=0.001 as reported in Figure 7.

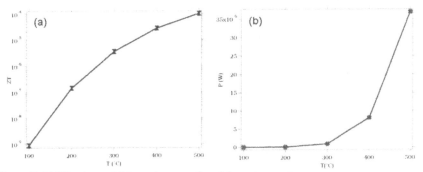

Figure 7. (a) ZT value and (b) ouput power of module vs. temperature.

CONCLUSIONS

Several thermoelectric modules were fabricated utilizing Al_2O_3 (2%, 5%, 8%) doped sintered ZnO (n-type) and pure sintered $Ca_3Co_4O_9$ (p-type) elements. Practical testing of the modules revealed that the best performance was obtained in the module formed by six 2 % Al_2O_3-ZnO/ $Ca_3Co_4O_9$ couples, that generated a power P=3.7 $\times 10^{-5}$ W and a voltage V = 300 mV at 500°C (when ΔT = 260°C) with a corresponding figure of merit ZT = 0.001. P and ZT values of our module are extremely small in comparison to results obtained by modules comprised of similar n and p materials [11], which number of elements was quite larger (140 couples vs. our 6).

At present, we are looking forward to enhance the performance of modules achieving a bigger figure of merit, especially by increasing the electrical conductivity of doped ZnO and $Ca_3Co_4O_9$ materials.

REFERENCES

1. "Harnessing Materials for Energy" V.S. Arunachalam and E.L. Fleischer, Eds., MRS Bull. 33(4) 2008 pp. 250-477.
2. "Harvesting Energy through Thermoelectrics: Power Generation and Cooling", T. M. Tritt and M. A. Subramanian, Eds., MRS Bull. 31(3) 2006 pp. 188-229.
3. G. J. Snyder and D. S. Toberer, Nature Materials 7, 105 (2008).
4. R. Venkatasubramanian, E. Siivola, Th. Colpitts and B. O'Quinn, Nature 413, 517 (2001).
5. B. Poudel, Q. Hao, Y. Ma, Y. Lan, A. Minnich, B. Yu, X. Yan, D. Wang, A. Muto, D.Vashaee, X. Chen, J. Liu, M. S. Dresselhaus, G. Chen, Z. Ren, Science 320, 634 (2008).
6. "Chemistry of the Elements" N. N. Greenwood and A. Earnshaw Eds, Pergamon Press Ed., Hong Kong 1989, p 1496
7. D. V. Malakhov, Inorg. Mat. 30, 1 (1994).
8. T. Tsubota, M. Ohtaki, K. Eguchi, H Arai, J. Mater. Chem., 7, 85 (1997).
9. M. Sano, S. Horii, I. Matsubara, R. Funahashi, M. Shikano, J. Shimoyama, K. Kishio, Japan. J. Appl. Phys. 42, L198 (2003).
10. K. Park, K. Y. Ko, W. S. Seo, W. S. Cho, J. G. Kim, J. Y. Kim, J. Eur. Ceram Soc. 27, 813 (2007).
11. R. Funahashi, M. Mikami, T. Mihara, S. Urata, N. Ando, J. Appl. Phys. 99, 066117 (2006).

AUTHOR INDEX

SUBJECT INDEX

Printed in the United States
By Bookmasters